Puppy Training Book for Beginners

By Emma Rose

Train in just 7 days

Acknowledgement

I'd like to take a moment to thank everyone who helped me write this book, "Puppy Training Book For Beginners." This project would not have been possible without the help, direction, and support of the following people.

First of all, I want to say a big thank you to my good friend Evelyn, who is a skilled dog trainer. This book wouldn't have been possible without her knowledge and help. Evelyn's commitment to the health and training of dogs is truly amazing, and her help has been very helpful as I've been writing. Evelyn, thank you for being willing to share your information and experiences with others.

Thank you to everyone who has picked up this book and put their time and faith in my work. This project was started because of your interest in dogs, your love for them, and your wish to give them the best care possible. I hope that the tips and information in these pages will help you train your puppy and build a strong, caring relationship with them.

I'm also thankful for the support, motivation, and understanding that my family and friends gave me while I was writing. Your faith in me and readiness to listen or help kept me going and gave me the inspiration I needed to finish this book.

I also want to thank the people at my distributor who thought this topic was important and helped make the book what it is now. Your professionalism, knowledge, and dedication to putting out high-quality material are really admirable.

Last but not least, I want to say thank you to all the dogs and puppies I've met over the years. Every wagging tail, wet nose, and playful spirit has taught me something important about patience, kindness, and the joy of love that doesn't depend on anything. This book is for all of our animal friends who make our lives so much happier.

Again, thank you to everyone who helped make this book possible. May it be a useful tool that helps puppies and their owners get along well and brings joy to many homes.

Thanks, from the bottom of my heart.
Emma Rose.

Table of Contents

Give paw

Introduction

Welcome to Puppy Training Book For Beginners - Train Your Puppy In just 7 days, the best way to transform your naughty fur ball into a well-behaved, loved friend. Get set to embark on a fun adventure through puppyhood, where you'll learn how to make a loving, peaceful bond that lasts a lifetime.

Inside the pages of this book, you'll find a treasure trove of practical techniques, valuable tips, and expert advice that will help you bring out your puppy's full potential. Say goodbye to shoes that have been chewed up and hello to a well-mannered dog that will make you feel proud.

Prepare to be amazed as you read each chapter, where we discuss in detail the most important parts of training a puppy. From figuring out how your puppy acts to learning the essential commands, we'll give you the knowledge and suggest tools to make training fun and rewarding for you and your furry friend.

Immerse yourself in a world of step-by-step instructions where we show you how to create a safe and comfortable environment, teach basic commands with a dash of fun and rewards, set up a routine that brings out the best, and handle common behavior problems with ease.

But there's more. In this fantastic book, you'll also learn how to socialize your puppy, both with people and other dogs. Find out how to teach your puppy good manners that will impress guests, and watch as your puppy grows into a charming and well-mannered friend.

Our tips for building resilience will help your puppy feel more confident and overcome their fears. As you advance in your training, you'll have the opportunity to observe their growth and witness them master impressive tricks and moves that will captivate everyone.

We know that a healthy puppy is a happy puppy, so we work hard to ensure they have a well-balanced diet and exercise routine that keeps them healthy and helps them live longer. You'll also learn how to care for and groom your puppy, making it the envy of all its four-legged friends.

But that's not all. Learn to read your puppy's subtle cues and body language to determine their needs. Lastly, let us show you how to make an unbreakable connection with your furry friend and make memories that will last a lifetime.

GO, BUDDY

Basic Questions Regarding the Book

Is it really possible to train a puppy in just 7 days?

Yes, it is possible to make significant progress in training a puppy in just 7 days. Even though mastery and perfection may take longer, the main goal of "Puppy Training Book For Beginners - Train Your Puppy In just 7 days" is to give you a structured program and effective techniques that show results quickly.

During this training period, you will set your puppy's behavior on a solid foundation, teach them the essential commands, and deal with common problems. The goal is to speed up the training process, teach your puppy good habits, and get them started on the way to becoming a well-mannered, obedient friend.

Within 7 days, you can see significant changes in your puppy's behavior if you invest consistent time and effort, follow the instructions, and use positive reinforcement techniques. But it's important to remember that these behaviors must be reinforced, practiced, and trained on beyond the first week if they are to be kept and improved over time.

Puppy Training Book For Beginners - Train Your Puppy In just 7 days takes a focused and condensed approach, applying the essential training ideas into a comprehensive guide that helps you maximize your time and speed up your puppy's learning. We want to give your puppy a good start by teaching the fundamentals of good behavior and paving the way for a lifetime of good relationships and happy companionship.

Is the "Puppy Training Book For Beginners - Train Your Puppy In just 7 days" program suitable for all breeds and sizes of puppies?

The "Puppy Training Book For Beginners - Train Your Puppy In just 7 days" plan has been designed so that all breeds and sizes of puppies can use it. The ideas and methods taught in the book can be used with any puppy, regardless of breed or size.

All puppies can learn from the basic ideas of positive reinforcement, consistency, socialization, and creating a safe environment. These rules are about understanding your puppy's behavior, setting clear expectations, and building a strong foundation of good manners and obedience.

Even though each puppy may have a different personality and way of learning, the core ideas of the program can be changed to fit the needs of different breeds and sizes. The book has a flexible approach that can be altered to fit the needs and characteristics of each puppy.

Whether your puppy is a small or large breed, the "Puppy Training Book For Beginners - Train Your Puppy In just 7 days" book gives you a complete plan to train your puppy. Following the rules and methods, you can prepare and shape your puppy's behavior well, regardless of breed or size.

So, you can rest assured that the program works with puppies of all breeds and sizes. This will help you train your furry friend well and build a strong bond with him or her.

What should I do if my puppy doesn't respond well to the training methods within the 7-day timeframe?

Even if your puppy doesn't respond as quickly as expected to the training methods in the book "Puppy Training Book For Beginners - Train Your Puppy In just 7 days," there's no need to worry. Every puppy is different; they all learn and grow at different rates.

The book "Puppy Training Book For Beginners - Train Your Puppy In just 7 days" gives a focused, intensive training plan that is meant to get good results quickly. But it's important to remember that each puppy learns at its own pace, and some may need more time and repetition to fully understand the training concepts.

If your puppy doesn't respond as quickly as you'd like, it's important to stay patient, positive, and consistent. Check your training methods to ensure you follow the book's guidelines correctly. Make sure that you are giving your puppy clear and consistent cues, that you are using positive reinforcement effectively, and that you are keeping his environment safe and interesting.

If you have problems or challenges during the training process, the book has more chapters and advice for you after the first 7 days. It gives much information about common behavior problems, advanced training methods, and keeping good habits for a long time.

Remember that training is a process that never ends, and some puppies may need more time to make progress. It's important to change the way you train your puppy to fit his or her needs, personality, and way of learning. Learn what their limitations are, and modify your training program accordingly.

The book "Puppy Training Book For Beginners - Train Your Puppy In just 7 days" is meant to give you the information and tools you need to train your dog well. If you stay committed, patient, and positive, you can keep working with your puppy after the first week and get the results you want over time.

So, be determined, keep trying, and enjoy the process of getting to know your puppy. The book gives you useful tools and tips to help you along the way, ensuring you and your furry friend can build a strong and peaceful relationship.

Understanding Your Puppy

The Developmental Stages of a Puppy:

To take good care of a dog, a new owner needs to know how puppies grow and change. Puppies go through different stages as they grow. Each stage has its own needs and traits. You can take good care of and train your puppy in a way that helps it grow and develop if you know about these stages.

The Neonatal Stage

The neonatal stage lasts about two weeks after birth. During this time, puppies need their mother for everything, including food and warmth.

1. The neonatal stage is a very important time for puppies. It starts right after they are born and usually lasts for about two weeks. During this time, the puppies depend on their mom. It's important to know that when puppies are born, they don't have eyes or ears yet, so

they can't see or hear. So, they use their senses of touch and smell a lot to find their way around and interact with the world around them.

2. The mother dog and her puppies must be in a peaceful and secure environment at this time because they depend on their mother to survive. Give the mother a good, quiet place to care for her babies. The puppies will be healthier and grow faster in this calm area. What they go through as puppies has a significant impact on how they grow up and behave as adults.

3. To make a safe and quiet space, you should limit noise, interruptions, and touching things too much. It's important to keep the area where the puppies live clean and free of anything that could hurt them. Also, keeping the right temperature is important because young puppies can't control their body temperature well. Making sure they have a warm, comfortable place to live keeps them healthy and helps them grow.

4. During the newborn stage, when the puppies are very young, the mother dog is important in caring for and feeding them. She gives them food, nurses them, and helps keep them warm and clean by grooming them. During this stage, the bond between the mother and her puppies is essential because it sets the stage for their future growth and socialization.

The Transitional Stage

Between the ages of two and four weeks, puppies undergo a significant transformation in their development.

1. With their newfound ability to see and hear, puppies become more aware of their surroundings and exhibit a growing curiosity. They start to pay closer attention to the sights and sounds around them, demonstrating an eagerness to explore and interact with their environment. This stage marks an important milestone in their development as they gradually transition from relying solely on their sense of touch and smell to utilizing their visual and auditory senses as well.

2. During this time, introducing the puppies to human touch and voices becomes highly beneficial. Gentle interactions with humans help them become familiar with human presence and handling, which is essential for their socialization. Regular and positive experiences with human touch contribute to their comfort and trust around people, fostering a healthy and positive relationship with humans as they grow older.

3. Let the puppies get used to being touched and hearing human voices. This helps them get used to different objects and learn to connect them with good things. This exposure helps them develop a better-rounded and adaptable temperament, making them better prepared for future interactions and relationships, including potential adoption into loving homes.

4. It's important to note that while introducing the puppies to human touch and voices, it should be done with utmost care and gentleness. Their senses are still developing, and they

may be sensitive to loud noises or sudden movements. Creating a calm and controlled environment during these interactions ensures that the puppies feel safe and comfortable while gradually becoming familiar with human presence.

The Socialization Stage

Between 4 and 12 weeks of age, puppies enter a crucial stage of their development where they need to learn how to interact and coexist with other individuals, both human and animal. This period is highly significant for their emotional and social growth. As puppies mature, they become increasingly interested in their surroundings, more playful, and more receptive to learning.

During this stage, it is essential to expose puppies to a variety of experiences and stimuli to help them become well-rounded and confident dogs. This includes introducing them to different people, animals, sounds, and environments. It is beneficial to expose them to individuals of different ages, appearances, and behaviors to ensure they are comfortable and adaptable in diverse social settings. Their capacity to develop trustworthy relationships with people throughout their lives is influenced by positive interactions with people during this stage.

a. Introducing puppies to other animals is also crucial during this stage. It allows them to learn how to communicate and coexist with their fellow furry companions. Proper socialization with other animals, including dogs of various sizes, breeds, and temperaments, helps puppies understand canine body language, establish appropriate boundaries, and develop valuable social skills. These experiences contribute to their ability to interact and play harmoniously with other dogs as they grow older.

b. Additionally, exposing puppies to different sounds and environments helps to become more adaptable and confident in new situations. Gradually introducing them to common household noises, such as vacuum cleaners or doorbells, as well as unfamiliar sounds and environments, prepares them to handle a variety of sensory stimuli throughout their lives. This exposure reduces the likelihood of developing fear or anxiety towards novel experiences, making them more resilient and well-adjusted dogs.

c. During this stage, it is important to prioritize positive and supervised interactions. Ensuring that the experiences are safe, controlled, and free from harm allows puppies to build positive associations and develop trust in their surroundings. Encouraging playfulness, exploration, and learning in a nurturing environment further supports their emotional and social development.

The Juvenile Stage

Puppies enter the juvenile stage between the ages of three and six months. Puppies go through a big change in the ages. During this time, puppies grow quickly both physically and mentally, making great strides in many areas of their development. Their coordination gets better, which makes them better at moving around and more able to perform on their own.

1. During the puppy stage, the puppies' bodies change in apparent ways. As their bones, muscles, and organs continue to develop, they get bigger and heavier quickly. Along with this growth spurt, their coordination, balance, and motor skills all improve. Puppies get faster and better at moving around their environment, showing that they have more control over how they move.

2. During the puppy stage, puppies also make significant changes in their minds. They become more interested, alert, and willing to learn. Their cognitive skills grow quickly, making it easier to understand and remember things. This is a great time to teach basic obedience skills and reinforce good behaviors.

3. Basic obedience training is essential for puppies to learn good habits that will stay with them into adulthood. It means teaching puppy's basic commands like "sit," "stay," and "come," as well as how to walk correctly on a leash and have good manners. Puppies learn to understand and follow commands with consistent training and positive reinforcement, like treats and praise. This builds a strong foundation for obedience and communication between the dog and its owner.

4. Also, training them at this age helps them learn how to get along with people and other animals. They learn essential skills for getting along with other people and dogs, like greeting and playing with them. Positive interactions with different people and situations help them build social skills and self-confidence.

5. A puppy's behavior and manners are also primarily shaped by how it acts when it's young. Consistent training and rewards for good behavior during this stage help set up good habits and discourage bad ones, like barking too much, jumping up and down, or chewing things up. Owners can help puppies learn self-control and make good decisions by setting clear limits, giving consistent guidance, and giving positive reinforcement.

It's important to remember that training and rewards for puppies should be based on how old they are and what they can do. Training sessions should be short, engaging, and fun to keep him from getting bored. At this stage, training needs to be done with patience, consistency, and positive reinforcement.

The Adolescent Stage

Puppies enter their teenage stage between the ages of six and eighteen months, and just like human teenagers, they may exhibit behaviors that can be challenging to handle. During this phase, puppies undergo various developmental changes, which can include becoming more independent, testing boundaries, and occasionally showing resistance to training. Patience, clear instructions, and consistent guidance are key factors in ensuring that the puppy matures into a well-behaved adult dog.

a. One characteristic of the teenage stage in puppies is an increased sense of independence; they start to explore their surroundings more autonomously and may display a desire for more freedom. It is important for owners to strike a balance between granting appropriate independence and maintaining necessary control. Allowing the supervised puppy opportunities to explore and make decisions within safe boundaries can help satisfy their growing independence while ensuring their safety and well-being.

b. Another common behavior during this stage is boundary testing. Puppies may challenge rules or push the limits set by their owners. This behavior is a normal part of their development as they strive to establish their own identity and understand their place in the hierarchy. Consistent enforcement of rules and clear boundaries is essential during this stage to establish the owner's authority and reinforce expectations. Positive reinforcement techniques, such as rewards for appropriate behavior, can be effective in encouraging the puppy to make good choices.

c. It is also not uncommon for puppies to go through temporary periods where they seem to forget their training or show resistance to it. This can be frustrating for owners who have previously seen progress in their puppy's behavior. However, it is important to remember that these regressions are usually temporary and a natural part of the developmental process. Consistency and patience in training are crucial during these periods. Revisiting and reinforcing previously learned commands, using positive reinforcement, and maintaining a calm and positive training environment can help the puppy overcome these challenges and continue progressing.

d. Clear instructions are vital during the teenage stage. Puppies need clear and consistent communication to understand what is expected of them. Using concise commands, gestures, and positive reinforcement can help ensure that the puppy comprehends and responds appropriately. Consistency in expectations and training methods across all family members also helps the puppy understand and follow instructions effectively.

Additionally, maintaining a calm and patient demeanor is crucial when dealing with teenage puppies. Frustration or anger can hinder the training process and strain the relationship between the owner and the puppy. Positive reinforcement, praise, and rewards for good behavior are more effective in shaping the puppy's behavior and promoting a strong bond between the owner and the dog.

The Adult Stage

After 18 months, your puppy will grow up and become an adult. Their behavior becomes more stable and easier to predict as they keep growing. At this point, you can improve their training, teach them more advanced commands, and fix any other behavior problems.

1. As puppies grow into adults, they have already learned the basics of obedience and good behavior. This foundation lays the groundwork for learning more advanced skills. It's important to remember that dogs, like people, keep learning and changing throughout their lives. This means training and changing a dog's behavior can still work even when the dog is an adult.

2. One part of training adults is teaching them more complicated commands. Starting with simple commands like "sit," "stay," and "come," dog owners can train their dogs in more complex orders and tasks. These may include "heel," "leave it," "fetch," or even more specific commands based on the dog's abilities or training goals. To teach these advanced commands, you must be consistent, use positive reinforcement, be patient, and do things repeatedly.

3. Along with learning commands, it's essential for adults to deal with any remaining behavior problems or issues. Some dogs may have long-term behavior problems that need to be fixed, like pulling on the leash, being afraid of being alone or reacting out of fear. Working with a professional dog behaviorist can help you determine why these behaviors are happening and develop effective ways to change and improve them.

4. Socialization can also help dogs who are already grown up. Dogs become well-adjusted and confident in different situations when they are exposed to other places, people, animals, and stimuli. Keeping up with positive, controlled socialization can help prevent or fix behavior problems caused by fear or aggression.

5. Even when a dog is an adult, it still needs exercise and mental stimulation. Regular physical activity, like daily walks, playtime, or interactive activities, helps the dog meet its material needs and can improve its overall health.

6. Even though an adult dog's behavior tends to be more stable, it is still important to reinforce good behavior and keep expectations the same. Positive reinforcement, like treats, praise, and affection, is an excellent way to get a dog to do what you want and strengthen the bond between you and your dog. Even as adults, consistency, patience, and understanding are still crucial for training and managing behavior well.

Determine a Schedule

Getting your puppy into a consistent routine means making a daily schedule for him or her. Here is a detailed explanation of each part of a schedule that needs to be thought about:

a. **Feeding Times:** Schedule when your dog will eat. Depending on their age and breed, most puppies need three to four meals daily. Divide the total amount of food for the day into the meals and set regular times to eat.

b. **Breaks to the bathroom:** Puppies have small bladders, so they must go to the bathroom often to avoid accidents. Take your puppy outside to use the bathroom right when it wakes up, after each meal, at playtime, during exercise, and before bedtime. Use consistent cues, like "Go potty," to get your dog to go to the bathroom. Be patient and treat your puppy well when it goes to the bathroom.

c. **Exercise/Play sessions:** Schedule times for your puppy to play and get some exercise. How much and what kind of exercise your puppy needs depends on age, breed, and energy level. Short walks, playtime with other puppies, and games that interest puppies are all great options. Give your puppy mental and physical activities to do during these sessions to help it get rid of excess energy and live a healthy life.

d. **Training Sessions:** Training is an important part of a puppy's daily schedule. Set aside time every day for training sessions. Focus on teaching simple commands, good manners, and how to get along with other people. Keep training sessions between 5 and 10 minutes to match your puppy's ability to pay attention. Use treats, praise, and rewards, among other things, to motivate and reinforce the behaviors you want.

e. **Resting:** Puppies need a lot of sleep and rest to help them grow and develop. Set aside specific times during the day for your puppy to rest. Set aside a quiet place for your puppy to rest and relax, like a crate or a cozy corner. Encourage them to rest during these times to keep them from getting too excited and to improve their overall health.

When making the schedule, it's important to think about your own schedule and obligations. Ensure that the schedule fits with the puppy and can be followed consistently.

Remember, consistency is crucial for establishing a routine. Dogs do best when they know what to expect and can plan for it. Follow the schedule as closely as possible, but be ready to make changes if your puppy needs them or if something unexpected arises.

Basic Canine Behavior

To build a strong relationship, you need to know how puppies generally act. Dogs have certain instincts and behaviors that are deeply rooted in their own nature. You can make your home a good place for your puppy to grow up if you know about and accept these behaviors.

Communication Signals

Dogs talk to each other through their body language, sounds, and facial expressions. Dogs show how they feel and what they want to do by wagging their tails, barking, growling, and moving their ears. Reading these signs will help you determine what your puppy wants, how he feels, and how comfortable he is in different situations.

a) Tail wagging is a common way for puppies and dogs to talk to each other. Many people think of a wagging tail as a sign of happiness, but it's important to remember that a wagging tail can mean different things in different situations. A comprehensive, relaxed wag usually means the puppy is friendly and happy, while a stiff or high-held tail could mean the puppy is alert, excited, or even angry. It's important to look at the puppy's body language and overall behavior to get a clear picture of what it's trying to say.

b) Dogs can also talk to each other by making sounds like barking, growling, and whimpering. Barking can be used to show excitement, warn of possible dangers, show anxiety, or get a dog's attention, among other things. Growling is usually a warning or a way to defend while whimpering or whining can be a sign of discomfort, fear, or a need for attention or help. Paying close attention to the context, intensity, and body language can tell you a lot about how your puppy is feeling.

c) Face movements are also a big part of how dogs talk to each other. Dogs can show many different feelings with their faces, like happiness, fear, anger, or submission. When calm and happy, the dog's eyes are usually open, soft, and relaxed. On the other hand, eyes that are narrowed, staring, or a tense face may show anger or discomfort. It's essential to look at your puppy's whole face, including the position of its ears, mouth, and body as a whole, to figure out how it's feeling.

d) Another important way a dog talks is through its ears. You can figure out a lot information about how they feel just by watching where and how they move their ears. Ears that stand up and face forward often mean that a person is paying attention, is curious, or is alert. Ears that hang down or fold back can be a sign of fear, submission, or relaxation.

To figure out what your puppy is trying to tell you, you need to look at the whole situation and consider body language, vocalizations, facial expressions, and ear position, among other things. Each dog is different, and while there are some common ways that dogs talk to each other, you should also consider individual differences and breed-specific traits. Spending time observing and getting to know your puppy's unique way of communicating will help you learn more about their needs, feelings, and comfort levels in different situations.

Pack Mentality

As pack animals, dogs are always looking for social orders and friends. By understanding this pack mentality, you can show your puppy you are the pack leader and build a strong relationship with him. To gain your puppy's trust as a leader, you must be consistent, give positive reinforcement, and talk to it.

Basic Instincts

Dogs have instincts that help them act the way they do. Some of these instincts are to hunt, protect, and defend. Even though these behaviors can be taught and controlled, it's important to remember that your puppy may still have some instincts. Understanding these instincts and dealing with them in the right way can help stop bad behaviors and make sure a dog is happy and well-adjusted.

Play Behavior

Play is one of the most important things for a puppy to do. It helps them learn how to get along with other people, stay in shape, and keep their minds active. Understanding the different ways that puppies play, like chasing, wrestling, or chewing, can help you give them the right opportunities to play and choose the right toys. Play also helps you and your puppy get closer, which makes for a good and fun relationship.

a. One of the best things about play is that it helps puppies to make friends and learn how to act in social situations. Puppies learn essential social skills like biting, barking, habitation, communication, and understanding boundaries. Puppies learn how to control their bite strength and improve their skills when they play with their littermates or other dogs. Additionally, playing helps them understand and respond to body language, which is necessary for getting along well with people and other animals.

b. Playing regularly helps them burn off extra energy, stay at a healthy weight, and build strong muscles and good coordination. Puppies can get exercise by running, chasing, and playing fetch. This keeps them from getting too fat and causing health problems. Exercise also helps them sleep better and keeps them from doing bad things when they're too excited.

c. A puppy's play routine should include both physical and mental activities. Games like hide-and-seek and puzzle toys that give out treats keep their minds active and test their ability to think. A puppy's mental health and ability can be improved by giving it a variety of toys and activities that encourage problem-solving and exploration.

d. It's important to use suitable toys for your puppy's enjoyment and safety. Toys should be durable, the right size, and be made of safe materials for chewing. Interactive toys that engage their senses, such as squeaky toys, puzzle toys, or toys with different textures, can be beneficial. Rotating your puppy's toys can help keep them interested and keep them from getting bored with the same ones.

During playtime, you can also get to know your puppy better and strengthen your relationship. Participating in their play sessions shows that you are a fun and positive person. Play interactive games like fetch, tug-of-war, or hide-and-seek to create a shared experience that builds trust and strengthens your connection. During training, play can also be used as a reward to reinforce good behavior and enhance the bond between you and your puppy.

Bonding with Your Puppy

It's important for your puppy's emotional health and growth that you form a strong bond with them. A connection that is deep and based on trust will lead to a lifetime of friendship and loyalty. Here are some ways to connect with your puppy in a meaningful way:

Spend Quality Time Together

Spend time with your puppy regularly and without interruptions. Take part in activities like brushing, training, playing, or just cuddling. This one-on-one time gives you both a chance to focus on each other and makes your relationship stronger.

Use Positive Reinforcement

Positive reinforcement is a very effective and helpful way to train your puppy and build a relationship based on trust and respect. It involves giving treats, praise, and love when your puppy does something good. This creates positive associations and encourages your puppy to do good things. This strengthens the bond between you and your furry friend by offering you a source of good things for them to experience. It also makes them more confident and trusting of you.

a. When you use positive reinforcement, it's essential to know exactly what you want your puppy to do. For example, if you want them to sit when commanded, you would focus on rewarding them when they successfully do it.

b. Using treats as a form of positive reinforcement works very well because a dog's natural drive is to get food. Choose treats that are tasty and fun for your puppy, like small pieces of soft dog treats or their favorite human foods that are safe for dogs. Since you reward your puppy with treats when it does what you want, it will be more likely to do it again in the future.

c. Along with treats, praise and affection are also important parts of positive reinforcement. Saying "Good job!" or "Well done!" positively and enthusiastically lets your puppy know they did something right. As a reward, you can also give your puppy a soft touch on the head, scratch behind the ears, or rub his belly. But it's important to remember that some dogs might not like certain kinds of physical contact, so paying attention to their preferences and acting accordingly is essential.

d. When using positive reinforcement, it is essential to be consistent. Giving your puppy immediate feedback and rewards is necessary when they do what you want them to do. This helps them understand the link between what they do and the good things that happen because of it. For your puppy to learn that the same behavior always gets a treat, you should use positive reinforcement in the same way every time.

e. Positive reinforcement not only strengthens your relationship with your puppy but also helps your puppy learn in a good way. You can create a safe and caring space for your puppy to learn and explore without fear or anxiety by rewarding good behaviors instead of punishing bad ones. This helps them feel good about themselves and trust you as their caretaker and teacher.

Positive reinforcement should be used with clear communication and proper training methods. It's essential to ensure your puppy knows what you want from him and that you give him clear instructions through consistent commands, cues, and training sessions. Positive reinforcement works best with other training methods, like clicker training or shaping behaviors, to teach new skills and behaviors.

Be Patient and Consistent

When training and playing with puppies, you have to be patient and consistent. If you want to build a strong relationship with your puppy, don't use harsh punishments or yell at them. Instead, be patient with your puppy as it learns and give clear, consistent instructions. This will make the puppy feel safe and help to trust each other.

Show Unconditional Love and Care

Your puppy needs love, care, and safety from you. Give them unconditional love and meet their physical and emotional needs. Being a responsible pet or animal owner means ensuring your pet gets enough exercise, eats well, sees the vet regularly, and gets groomed as needed. Your love and care for your puppy will make the bond between you two stronger.

a. Your constant love for your furry friend is the foundation of a strong bond between you and it. For their overall health and happiness, it is essential to meet their physical and emotional needs. You can ensure your puppy gets enough exercise, good food, medical care, and grooming to give them a safe and happy home.

b. Your love for them should be unconditional, showing that you care about them no matter what they do or how they act. Spending quality time with your puppy, playing, cuddling, and training, strengthens your relationship with him. Your love makes them feel valued and loved when you give them a sense of safety and belonging.

c. It's important for your puppy's health and well-being that you meet their physical needs. Regular exercise is essential to keep them in good shape, keep them at a healthy weight, and give them more energy. Activities like walks, playing, and interactive games help puppies burn off energy, keep their minds active, and avoid behavior problems. Make sure your puppy gets the right amount of exercise by adapting their exercise routine to their breed, age, and energy level.

d. Proper nutrition is also important for the health and growth of your puppy. Talk to your vet to determine the best diet for your puppy, including the right balance of nutrients and the right amount of food. Give them good food that will help them grow, keep their weight steady, and keep their skin and coat healthy. Make sure they can always get fresh water to stay hydrated.

e. Schedule regular checkups, vaccinations, and preventive treatments to keep common diseases and parasites from hurting your puppy. Any health issues can be identified and treated early when your pet receives routine checkups from the veterinarian. Follow your vet's instructions about immunizations, deworming, flea and tick prevention, and dental care to make sure your puppy is healthy in the long run.

f. Taking care of your puppy's emotional well-being is as important as meeting its physical needs. Make sure they have a lot of ways to think and talk to other people. Give them training sessions to keep their minds active and boost their self-esteem. Introduce them to other animals, people, and places so they can grow up to be well-adjusted, confident dogs.

It is essential to make your puppy's environment safe. Make your home safe for puppies by eliminating any dangers or toxic substances that could hurt them. Give them a safe and comfortable place to sleep, like a crate or a bed, where they won't be bothered. Keep them away from dangerous things like extreme weather or mean animals, and watch them when they are outside or in a new place.

Choosing the Right Puppy Training Method

Choosing the right way to train your puppy is important if you want the training to be effective and kind and to build a strong bond with your furry friend. There are many ways to train, and each one has its own approach and way of thinking.

1. **Positive reinforcement:** Training methods that use positive reinforcement focus on rewarding good behaviors instead of punishing or correcting bad ones. Using treats, praise, toys, or other rewards to encourage and reinforce good behavior is part of this method. Positive reinforcement makes your puppy feel good about training, builds its confidence, and strengthens the bond between you and your puppy.

2. **Force-Free and Gentle Techniques:** Choose training methods that don't use force and are gentle. Stay away from strategies that use physical punishment, intimidation, or fear. Positive training methods make your puppy feel safe and cared for, so the puppy can learn without fear or stress. These methods help you build a relationship with your puppy based on trust and cooperation.

3. **Science-Based Approaches:** Think about training methods that are based on scientific research and a good understanding of how dogs behave. Look for methods of training that have been proven to work and that consider how animals learn, like classical conditioning and operant conditioning. Science-based methods ensure the training is effective and matches your puppy's natural instincts and cognitive abilities.

4. **Individualized Approach:** Choose a way to train your puppy that can be changed to fit its needs, personality, and way of learning. Every puppy is different, so a one-size-fits-all method might not work. Better results will come from a training method that can be changed and adapted to fit your puppy's personality.

5. **Clear Communication:** Ensure the training method emphasizes clear and consistent communication between you and your puppy. To talk to your puppy clearly, use cues,

signals, and body language that it can easily understand. The method should teach you to be patient, pay attention, and react to your puppy's needs and cues.

6. **Long-Term Changes in Behavior:** Consider whether the training method is geared toward long-term changes in behavior or just short-term obedience. It's important to teach your puppy the basic commands and how to act and make decisions. A complete training method helps your puppy learn good manners and self-control and make the right decisions.

7. **Professional Advice and Help:** If you aren't sure about the best way to train your dog or face problems, talk to a professional dog trainer or behaviorist for advice. They can help you choose and use the best way to train your puppy by using their knowledge, personalized advice, and hands-on support.

8. **Consistency and patience:** No matter what training method you choose, the keys to success are consistency and patience. Use the training techniques and cues consistently, and be patient with your puppy as it learns. Training takes time, and your progress may be slow at first. A method emphasizing patience and consistency will help you and your puppy stay motivated and get long-term results.

Remember that building a solid relationship and trust with your dog is just as crucial as teaching it commands, and the right way to train it will help you do both.

Addressing Barking and Excessive Vocalization:

Identify the cause: Find out why your dog is barking so much. It could be because of boredom, fear, anxiety, wanting attention, being territorial, or a health problem. Make sure your puppy gets regular exercise, playtime, and mental stimulation with interactive toys or puzzle games.

1. **Teach your puppy the word "quiet":** To teach your puppy the "quiet" command, you should use training methods that give him or her positive feedback. Reward them when they listen to you and stop barking.

2. **Counter-conditioning and desensitization:** Expose your puppy gradually to the things that make them bark too much, and use positive reinforcement to make good associations and stop the barking.

Talk to a professional dog trainer or behaviorist for help if the barking doesn't stop or is causing a lot of stress.

How to deal with separation anxiety and crate problems:

1. **Gradual desensitization:** Get your dog used to being alone by leaving him or her alone for shorter and shorter amounts of time at first and then gradually increasing the time. Use positive reinforcement and give them treats or toys when they stay calm.

2. **Crate training:** Offer treats, toys, and comfortable bedding inside the crate to make it a good experience. Gradually lengthen the amount of time the dog stays in the crate while giving treats for calm behavior.

3. **Set up a routine:** Set up a regular daily schedule that includes exercise, feeding, and time alone. Dogs do best when they know what to expect, and having a set routine can help them feel less anxious.

4. **Provide mental stimulation:** Leave puzzles, interactive toys, or frozen Kong toys filled with food to keep your puppy's mind busy and off of you while you're gone.

How to Deal with Aggression and Reactivity on the Leash:

1. **Identify triggers:** Determine what triggers your dog's reactive or aggressive behavior. It could be other dogs, strange people, strange places, or certain things.

2. **Training based on positive feedback:** Use positive reinforcement methods to reward calm behavior that doesn't cause trouble. Gradually expose your puppy to the triggers from a safe distance and give them treats when they stay calm.

3. **Avoid punishment:** Punishment can make puppies more afraid and angrier. Instead, use training methods that are based on positive reinforcement and rewards.

4. **Handling the leash:** Use a harness or head collar that fits your dog well to have more control over him when you walk with him. Keep a safe distance from triggers, and don't pull on the leash too hard, which can make the dog more reactive.

Consult a dog trainer or behaviorist specializing in aggression and reactivity if your dog's aggression or reactivity to being on a leash is severe, unpredictable, or dangerous.

Preparing for Training

Creating a Puppy-Friendly Environment:

Getting the puppy used to its new home is important for training to go well. You can help your puppy learn and get ready for training by giving them a safe and fun place to play.

Puppy-Proofing

Before you bring your puppy home, you need to make sure that your home is safe for puppies. Puppies are naturally curious and often use their mouths to find out more. Get rid of anything that could hurt, like poisonous plants, chemicals, small objects, electrical cords, and things that are easily broken. Don't let animals chew on loose wires and cables. This will make sure your puppy is safe and reduce the chance of accidents.

1. **Remove Potential Dangers:** Look around your house carefully to see if any things could hurt your puppy. This includes things they can reach, like poisonous plants, cleaning supplies, chemicals, medicines, and household items. Keep them out of your puppy's reach by putting them in cabinets or on high shelves.

2. **Secure Loose Wires and Cords:** Puppies may chew on loose wires and cords, hurting them or giving them an electric shock. Use cord protectors or cable organizers to bundle and secure loose wires or cords. You can also use bitter sprays or safe deterrents to stop your puppy from chewing on them.

3. **Store small things:** Puppies tend to find and swallow small things, which can cause them to choke or get a blockage in their intestines. Keep your puppy away from small items like coins, buttons, jewelry, and kids' toys. Watch out for hairpins, rubber bands, and sewing supplies that could fall on the floor.

4. **Secure Trash and Recycling Bins:** Make sure your trash and recycling bins have secure lids or are stored in cabinets your puppy can't reach. Get rid of things that could hurt, like sharp objects, toxic substances, or food scraps that could hurt.

5. **Protect Things That Are Easily Broken:** Puppies are known for having a lot of energy, and their tails can accidentally knock over or break things like vases, lamps, or picture frames. Consider putting fragile or expensive items on higher shelves or closed cabinets.

6. **Close-Off Restricted Areas:** Use baby gates or other barriers to prevent your puppy from entering certain parts of your home. This could be a room with fragile furniture, dangerous materials, or a place you want to keep out of reach for safety reasons.

7. **Watch Your Puppy:** Even if you've done an excellent job of puppy-proofing your home, watching your puppy when it's out exploring is essential. Keep an eye on what they are doing and get their attention away from something dangerous if they seem interested in it.

Designated Puppy Area

Set up a place for your puppy to chill out, sleep, and play. This area should be easy to get to and comfortable. Give them a cozy bed or crate to go to when they need to sleep. Use baby gates or playpens to create boundaries and limit access to certain parts of your home until your puppy is fully trained.

1. **Choose the Right Place:** Choose a place in your house that's good for your puppy. It should be a quiet place away from lots of people, loud noises, and a lot of other activities. Think about a site that is easy for your puppy to get to and lets you keep an eye on what they are doing.

2. **Give your puppy a cozy bed or crate:** Your puppy needs a comfortable place to sleep and rest. Invest in a bed or crate that is well-padded and the right size for them. This will make them feel safe. Choose a bed or crate made of materials that are easy to clean and will last a long time.

3. **Use baby gates or playpens:** Baby gates or playpens can help you set up boundaries and limit where your puppy can go in your home. They let you give your puppy a defined space while giving it some freedom to move around. This is especially helpful at the beginning of house training when you want to limit their access to certain rooms or areas.

4. **Include Important Things:** Ensure your puppy's designated area has all the essential things they need for their health. This could include their food and water bowls, chew toys, interactive toys, and a place to go to the bathroom (if they use puppy pads or a litter box). You and your puppy will find it easier to use these things when they are in their designated spot.

5. **Think about a Play Area:** Depending on your space, you may want to set up a separate play area within the puppy area. This can be a small playpen or a part of the room where your puppy can play while you watch. Include puzzles, balls, and toys they can play with to keep their minds active and entertained.

6. **Give Your Puppy Enough Supervision:** Give your puppy enough supervision while it's in its designated area. Check on them often to ensure they have fresh water, clean, comfortable bedding, and the right toys and activities to keep them busy. You can also help your puppy if it needs it or if you see something that could be dangerous.

As your puppy gets older and better at going to the bathroom and behaving, you can give them more access to other parts of your home. This should be done in a controlled way that only lets them into one room or area at a time. This way, they can still use their safe space.

Socialization Opportunities

Give your puppy a chance to see, hear, and do different things in a safe environment. Show them different things, surfaces, and textures. This helps them learn about the world around them and gives them more confidence. Give toys, puzzles, and enrichment activities that make them think and keep them from getting bored.

Quiet Retreat Spaces

Like people, puppies need quiet places where they can rest and re-energize. Give your puppy a separate room or crate where they can go when they are feeling overwhelmed or need some time to themselves. This gives them a safe place to relax and lowers the chance that they will have problems with anxiety or stress.

Important Training Equipment:

For training sessions to be effective and efficient, they need to have the right tools. Here are some essential supplies you'll need:

Leash and Collar

A collar and leash that fit well are important for teaching your puppy to walk on a leash and keeping them safe when you go outside. Choose a collar made of strong material, and if your puppy pulls or has a sensitive neck, you might want to use a harness instead.

Rewarding and Treating

During training, treats and rewards with a high value are powerful ways to keep people interested. Use small, soft treats that your puppy just can't say no to. You can also give them their regular food as a reward. Try out different treats to see what gets your puppy excited the most.

Clicker or Marker

To mark good behavior, you can use a clicker or a verbal cue, like the word "Yes!" or a clear sound. The clicker or marker tells your puppy that it has done something good and will soon get a treat. When you use the marker, in the same way every time, your puppy learns what behaviors you are rewarding.

Training Toys

You can use training toys like ones that give out treats or puzzles to make your puppy's training sessions more fun and interesting. These toys keep your puppy's mind active and help him focus on something else while you train him.

Training Mat or Blanket

Set aside a specific training mat or blanket as the place where you'll train your puppy. This gives them a clear place to focus on learning and helps set limits while they are being trained.

Establishing a Routine

Setting up a routine is important for your puppy's training and health as a whole. Dogs like things to stay the same and be predictable, and a well-structured routine can make them feel safe.

Day 1: Building a Strong Foundation

Introducing Your Puppy to Their New Home:

When you bring it home, it's important to make your home a happy and welcoming place for your puppy. Here's how to get your puppy used to their new home:

Set up a space for your puppy that is both safe and comfortable. This can be a small room, a crate, or a playpen. Ensure the area is safe for the puppy by eliminating any potential dangers.

Let your puppy look around in the new area at its own pace. Make sure to keep a close eye on them so they don't do anything they shouldn't. Give them praise and treats when they are curious and don't get into trouble while exploring.

Show your puppy where to find its food and water bowls, bed, toys, and the place where it is supposed to go potty. Make sure that these things are easy for them to find out.

Spend quality time with your puppy in their new space to help them. Sit or lie on the floor so they can come up to you and get used to your presence. To build trust, give gentle pats and positive feedback.

Set up a routine: Start a daily routine for feeding, going to the bathroom, playing, and sleeping. Your puppy will get used to its new home faster if you are consistent.

When you bring a new puppy home, it's important to ensure they feel comfortable and safe. One of the first things to do is show the puppy where all of its essential things are, like its food and water bowls, bed, toys, and the place where it is supposed to go potty. Put these things in easy-to-reach places so the puppy can find them without much trouble. This helps the puppy figure out where its basic needs can be met.

1. Setting up a routine is essential for the puppy's overall health and for helping it adjust quickly to its new home. Set up a daily schedule for feeding, going to the bathroom, playing, and sleeping. Consistency is important because puppies do best with routines and things they can count on. The puppy learns what to expect and when by sticking to a set schedule. This can make the puppy feel less anxious and more at ease.

2. It is critical to provide food on a consistent basis, preferably at the same time every day. This helps the puppy get used to eating simultaneously every day and helps with potty training since they will probably need to go potty soon after eating. Take the puppy to their designated potty area every time after they eat, and praise them when they go to the right place.

3. Puppies need to play to keep their minds and bodies active. Play with the puppy using suitable toys to encourage it to move around and have fun. This helps them eliminate their energy and keeps them from getting bored or doing something terrible. Make sure the toys are safe and suitable for the size and age of the puppy.

4. Maintaining a regular sleep schedule is essential for the puppy's health and growth. Make sure they have a comfortable place to sleep, whether it's a crate, a dog bed, or a particular spot in your home. To get the puppy to use this space, put soft bedding there and make it quiet and calm.

Always remember patience, consistency, and positive reinforcement are the most important things during this process. The puppy may need some time to adjust to their new surroundings, but with your help and attention, they will gradually grow accustomed to and comfortable in their new home.

The Basics of House Training:

House training is an important part of training a puppy. Here's what you need to know to get started:

1. Take your puppy outside to go to the bathroom often, especially after meals, naps, playtime, and waking up.

2. If you want to house-train your puppy, you need to set a schedule for when it needs to go potty. You can prevent accidents inside and assist them in developing good bathroom habits by giving them regular opportunities to relieve themselves in the proper location.

3. Puppies need to go outside often because their bladders aren't fully developed, and they can't control their bowel movements yet. Aim for every 1-2 hours at first, then modify the frequency based on the age and breed of your puppy. As they get older and can hold it for longer periods, gradually extend the intervals between bathroom breaks.

4. Puppies typically need to use toilets shortly after eating. After each meal, take your puppy outside for 10 to 15 minutes so they can relieve themselves. This strengthens the link between eating out and avoiding accidents inside the house.

5. When your puppy wakes up from a nap, they frequently need to go potty. Take them outside to their designated bathroom spot right away so they can relieve themselves.

6. Physical activity can increase the urge to use the restroom. Bring your puppy outside after a fun play session to give them a chance to go potty before returning inside. This helps to avoid accidents during or right after playtime.

Make it a habit to take your puppy outside as soon as they wake up, whether it's in the morning or after a nap. This helps them comprehend the desired behavior by reinforcing that going potty outside is the proper place.

Watching: Keep a close eye on your puppy, especially as they learn how to use the bathroom. Look for indications they need to relieve themselves, such as restlessness, sniffing, circling, or squatting. To prevent accidents inside, take them outside to their designated bathroom as soon as you see any of these signs.

Avoid Accidents: Accidents can occur during the house-training process, despite your best efforts. If you find your puppy having an indoor accident, gently stop them with a sound, like clapping your hands, and take them right outside to their designated potty area. As this may induce fear or anxiety in the puppy, refrain from reprimanding or punishing it. Instead, reward them for using the proper restroom by reinforcing the desired behavior.

The potty schedule must be established: You can lessen the possibility of accidents occurring inside by taking your puppy outside after meals, naps, playtimes, and when it first wakes up. Your puppy will eventually learn to associate using the restroom with the designated outdoor area, thanks to persistence, consistency, and positive reinforcement.

Positive reinforcement for potty schedule: Positive reinforcement is a great way to teach your puppy good behavior and get it to go to the bathroom in the right place. When your puppy goes to the bathroom in the right place, it's important to praise and reward it right away to keep this good behavior. Here's a detailed explanation of how positive reinforcement works and how to use it:

1. **Timing is important:** After your puppy goes to the bathroom in the right place, you should praise him, give him treats, or tell him he did a good job right away. The more closely the reward is related to the behavior you want them to do, the better they will connect the two.

2. **Verbal praise:** When your puppy goes to the bathroom in the right spot, use a cheerful and happy voice to show your happiness. Tell them they did something right by saying, "Good job!" or "Well done!" This helps them understand that they've made you happy and reinforces the behavior you want to see more of.

3. **Treats:** Puppies can be very motivated when you reward them with a small, tasty treat. When you take your puppy outside to the bathroom, have some treats ready. When they're done going to the bathroom, please give them a treat and praise them verbally. This makes the dog link going to the toilet in the right place with getting a tasty treat.

4. **Gentle petting:** Besides verbal praise and treats, gentle petting can also be an excellent way to show your dog that you like him or her. When your puppy goes to the bathroom in the right place, please give it some loving pats or strokes as a reward. Touching your puppy helps you get closer to it and reinforces how good it feels to go to the bathroom in the right place.

5. **Avoid punishment:** It's important not to yell at or hurt your puppy if it has an accident inside the house. Fear and anxiety can be caused by negative reinforcement, which can slow down the process of potty training and hurt the trust between you and your puppy. Focus on positive reinforcement instead to encourage good behavior and help them think of their new home well.

6. **Using praise, treats, and gentle petting:** You can create a positive environment for your puppy that encourages them to behave well and helps them remember where to go to the bathroom. This good memory builds trust between you and your puppy and makes them

feel good about their new home. Remember to be consistent and patient, and give positive reinforcement to build a creative relationship with your puppy and help them learn how to use the potty.

When your puppy needs to go potty, he or she will show you by sniffing the ground, circling, or whining. When you see these signs, you should quickly take them outside.

Puppy's exhibit specific behaviors and signs, known as "going potty," when relieving themselves. If you know what these signs mean, you can quickly take your puppy outside to the right place. Here's a full list of the signs to look out for:

1. One of the most common signs that a puppy needs to go potty is when it starts sniffing the ground. When a puppy sniffs the floor or ground in a specific spot, it's usually because they're looking for an excellent place to go potty. Dogs do this naturally because they use their sense of smell to find good places for the bathroom.

2. A puppy that starts to circle in a small area is another sign. This is often what they do right before they have to go potty. The circling helps them find a comfortable spot and makes them want to go to the bathroom. If your puppy keeps going around the same spot, it's a good sign that they must go outside immediately.

3. Puppies may make noises like whining, whimpering, or even barking when they need to go potty. This can be a way for them to immediately show they are uncomfortable or need

help. Some puppies may also be restless or irritable when they need to go to the bathroom. They may have trouble getting comfortable or moving around a lot, which is a sign that they need to go outside.

4. If you see any of these signs, you need to act fast and take your puppy outside to a good place for it to go potty. This makes them more likely to go to the bathroom outside and helps them learn how to do it right. If they wait too long to go out, they might have accidents inside, which can confuse them and slow their progress with potty training.

5. Keep a close eye on your puppy when it is inside, and keep it in a small area. When you can't watch them right away, use baby gates or a crate to keep them in a small, puppy-proofed place.

How to help your puppy get used to its new home

Bringing a puppy at home is exciting, but it can also be overwhelming for them. Here are some ways to make the move easier and more pleasant:

Set up a specific place: Set up a safe place for your puppy to feel at ease. Add a soft bed, toys, and bowls for food and water. This place will become a safe place for them.

Gradual exploration: Give your puppy time to get used to its new environment at its own speed. Start with a small space and slowly let them into more and more of your home. Watch them as they explore to make sure they stay safe.

1. When you bring a new puppy into your home, it's important to provide them with a secure location where they can feel comfortable. This area will be their safe place and help them adjust to their new surroundings.

2. Search for a good spot in your home where your puppy can have its own space. This area should be quiet, easy to get to, and far from anything risky. It could be a part of a room or a whole different room. Place a soft, cozy bed in the area so your puppy can rest and sleep in comfort. Also, give them bowls for food and water, as well as some toys to keep them busy. These familiar things will help them feel safe and make the space feel like theirs.

3. You should ease your puppy into their new environment so they don't get too stressed out. Start by putting them in a smaller space, like the spot you've set aside for them. You can keep them from entering other houses using a baby gate or a puppy pen. This small space makes them feel safe while they get used to their new environment.

4. As your puppy gets used to their designated area, you can give them more freedom by letting them explore more of your home. Open the gate or make the puppy pen bigger so they can access a bit more space. During this time of exploring, keep a close eye on them to ensure they are safe and avoid any accidents or damage to your home. Watch how they act and what they say to figure out how comfortable they are.

5. When your puppy is out exploring, you must keep a close eye on them to ensure they are safe. Puppies are naturally curious, so they may look at things, furniture, or places that could be dangerous. Watch out for something that could hurt them, like loose wires, poisonous plants, or small things they could swallow. Puppy-proof the places they can go by taking away or securing anything dangerous to their health.

6. Making their boundaries progressively more prominent, your puppy can explore and get used to your home at their own pace. This method helps them feel better about themselves and get used to their new surroundings while reducing the chances of accidents or anxiety.

Remember to give praise, rewards, and positive reinforcement when your puppy acts well while exploring. This makes them more likely to think of their safe place and slow exploration as good things, which adds to their sense of security.

Basics of Housetraining:

Housetraining is important for both your puppy's health and the cleanliness of your home. Here's a step-by-step guide to getting started:

Puppies do best with a regular schedule for eating, going to the bathroom, and playing. This helps them get into a routine and makes it easier to train them to go to the bathroom outside.

1. **Provide a secure location:** When you bring a new puppy into your home, it's important to provide them with a secure location where they can feel comfortable. This area will be a safe place and help them adjust to their new surroundings.

2. **Set up a safe place:** Search for a good spot in your home where your puppy can have its own space. This area should be quiet, easy to get to, and far from anything risky. It could be a part of a room or a whole different room. Place a soft, cozy bed in the area so your puppy can rest and sleep in comfort. Also, give them bowls for food and water, as well as some toys to keep them busy. These familiar things will help them feel safe and make the space feel like theirs.

3. **Slowly give them more freedom:** As your puppy gets used to their designated area, you can give them more freedom by letting them explore more of your home. Open the gate or make the puppy pen bigger so they can access a bit more space. During this time of exploring, keep a close eye on them to ensure they are safe and avoid any accidents or damage to your home. Watch how they act and what they say to figure out how comfortable they are.

4. **Watch out for safety:** When your puppy is out exploring, you must keep a close eye on them to ensure they are safe. Puppies are naturally curious, so they may look at things, furniture, or places that could be dangerous. Watch out for something that could hurt them, like loose wires, poisonous plants, or small things they could swallow. Remove or lock up anything that could hurt your puppy before letting them go to certain places.

5. **Boundaries:** Increasing the visibility of their boundaries will allow your puppy to explore and get used to your home at their own pace. This method helps them feel better about themselves and get used to their new surroundings while reducing the chances of accidents or anxiety.

6. **Give rewards**: Remember to give praise, rewards, and positive reinforcement when your puppy acts well while exploring. This makes them more likely to think of their safe place and slow exploration as good things, which adds to their sense of security.

Scheduled trips to the bathroom:

Starting a routine will help your puppy learn what to expect and when to do everything throughout the day. This helps puppies feel less stressed and anxious because they like structure and consistency.

Set regular times to feed your puppy, and always do it at specific times. Puppies thrive on routine, so making sure they eat at regular times helps their digestion and keeps them from feeling too hungry or full.

Also, having a routine is very important for potty training. If you feed your puppy at the same time every day, you can predict when it will need to go to the bathroom. This lets you plan their bathroom breaks and reinforce the habit of going outside.

1. When your puppy wakes up in the morning or after a nap, you should take it outside quickly. Puppies usually need to go to the bathroom soon after they wake up. Take your puppy to the place where it is supposed to go potty and let it go. Wait for them and give them enough time to do what they need to do.

2. Puppies usually have to go to the bathroom soon after they eat. Take your puppy outside to go to the toilet about 15 to 30 minutes after they finish eating. This gives their digestive system time to start working, which makes it more likely that they will need to go to the bathroom. Tell them to do their business by saying something like "Go potty" or "Do your business."

3. Active playtime can make a puppy want to go to the bathroom after a session. After a fun game, you should take your puppy outside so it can go to the bathroom. Play with them for a while, and then take them to where they are supposed to go potty. This routine helps your puppy learn to go to the bathroom outside by reinforcing the link between playtime and bathroom breaks.

4. Before it's time for bed, take your puppy outside for one last trip to the bathroom. This makes sure that they have gone to the bathroom before bed, which makes it less likely that they will have an accident during the night. Establishing a bedtime routine for your puppy will help them understand that they must use the toilets before falling asleep.

5. When teaching your puppy to go to the bathroom regularly, consistency's essential. Try to take bathroom breaks at the exact times every day. When you take your puppy outside, it learns that going to the bathroom outside is expected.

Recognizing names and simple commands:

Teaching your puppy its name and a few basic commands sets the stage for good communication. Here's what to do first:

Use your puppy's name often and in a positive way to help it learn it. When they look at you or answer you, give them treats and praise. This helps them connect their name to good things and makes them more likely to respond.

32

- **Sit command:** Move a treat up and down in front of your puppy's nose to teach it to sit. As their head moves toward the treat, their back end will naturally drop until they are sitting. Reward them right away, and then do it again, this time telling them to "sit."

- **Command:** "Stay." Start by telling your puppy to sit. Once they are sitting down, hold your hand up with the palm facing them and say "stay" in a firm but calm voice. Take a step back and reward them if they don't move. Gradually make the distance and length of the stay longer and longer.

Start by calling your dog inside with a short leash. Back away from your puppy while calling its name in a happy voice. When they come to you, you should praise them and give them treats. Practice this in different places to help them remember it better.

Day 2: Socialization and Communication

Puppy socialization is essential for their overall growth and health. Positive interactions with both people and other animals help them feel better about themselves, learn how to act properly, and make friends. Understanding and responding to your puppy's body language is a key part of being able to talk to him or her well. Let's talk in-depth about each of these topics:

Proper Puppy Socialization:

To socialize with your puppy, you should introduce them to different places, people, animals, and experiences safely and positively. This helps your puppy feel comfortable and confident in different situations, which makes it less likely that he or she will be afraid or aggressive as an adult.

Early socialization: Start getting your puppy used to other people and animals as soon as possible, ideally between the ages of 3 and 14 weeks. During this critical time, puppies are more accessible to new things and more likely to adapt perfectly.

Controlled exposure: Take your puppy to different places, like parks, streets, pet-friendly stores, and separate rooms inside other buildings. Gradually add more stimuli while making sure your puppy stays calm and happy. Use treats, toys, and positive reinforcement to help your dog think of new things in a good way.

1. **Start with places that are familiar to your puppy**: To start, take your puppy to new places that are calm and familiar to them. This could be different rooms in your own home or the homes of trusted friends or family. Let your puppy explore these areas at its own pace, and use treats, toys, and positive feedback to encourage it.

2. **Gradually add more stimuli:** Once your puppy is used to being in familiar places, slowly take them to places with more things going on. You could take them to a quiet park or a less busy street. The goal is to let your puppy see, hear, and smell new things without overwhelming them. Watch their body language to make sure they stay happy and calm.

3. **Controlled exposure:** When taking your puppy to new places, it's important to keep control over the environment and the amount of stimulation. Especially in public places, keep your puppy on a leash or in a carrier if you have to. This lets you lead them and keep them safe as they explore. Gradually give them more freedom as they get used to you and pay more attention to your cues.

4. **Use treats, toys, and positive reinforcement**: Use treats, toys, and positive reinforcement to make new places and experiences feel good. When your puppy meets something new and responds calmly and well, give it treats, praise, and lots of love. This helps them connect to new places with good memories, which boosts their confidence.

5. **Pay close attention:** Pay close attention to your puppy's body language and behavior to look for signs of stress. Stress or discomfort can show up as trembling, over-breathing, cowering, or trying to hide. If your puppy shows signs of stress, remove them from the situation and give them a break in a place they know and are comfortable with. Slowly bring back the stimuli one at a time.

6. **Be patient and consistent:** Socializing is an ongoing process that requires patience and consistency. Make it a habit to take your puppy to new places often and slowly increase the amount of stimulation. Each experience should be good for your puppy and teach them something new, which will boost their confidence and ability to adapt.

7. **Positive interactions:** Help people and animals have good interactions with each other. Set up play dates with well-behaved, vaccinated dogs so your puppy can learn how to behave around other dogs. Watch these interactions to ensure they are safe, and step in if anyone seems uncomfortable or angry.

Handling gently: Get your puppy used to being gently touched and held. Touch their paws, ears, and bodies gently so they get used to being touched by people. This helps keep the animal from getting scared or angry during routine veterinary visits, grooming, and other situations where it needs to be handled.

1. **Start with something you know:** Start by touching your puppy on its back or sides, where they are most comfortable. Use light pressure and gentle strokes. Watch how your puppy reacts to make sure they are calm and open to being touched.

2. **Gradually touch their paws, ears, and tail:** As your puppy gets used to gentle touch, you can start to touch their more sensitive areas, like their paws, ears, and tail. Slowly and gently approach these areas and use positive reinforcement techniques. For example, give them a treat or praise while lightly touching their paws. This helps your puppy connect being touched with good things.

3. **Positive association:** When your puppy responds calmly and positively to gentle touch, reward them with treats, praise, or play. This will help them form a positive association with touch. This makes it clear that touching is a good and rewarding thing to do.

4. **Consistency and repetition:** Touch your puppy gently every time you speak to or play with him. Be gentle with them every day. For example, pet them when you play with them or rub their bellies. Your puppy will get used to being touched and held if you are consistent and do it over and over again.

5. **Gradual progression:** As your puppy gets used to the touch, gradually lengthen it and make it stronger. But always be aware of their limits and try not to overwhelm them. Pay attention to their body language and cues to make sure they are not showing signs of stress or discomfort.

6. **Desensitization exercises:** You could do desensitization exercises with your puppy to help him or her get used to being touched. This means letting them slowly get used to different kinds of touch, like putting light pressure on their paws or ears. Start with soft touches and gradually add more pressure over time, ensuring to check how comfortable they are.

Socialization is an ongoing process as long as your puppy lives. Keep introducing them to new things, people, and animals. Reward good behavior and deal with any signs of fear or anxiety.

Positive Interactions with Humans and Other Animals:

Your puppy's social development depends on his good experiences with people and other animals. Here's how to make sure people get along:

1. **Encourage social greetings:** Teach your puppy to be calm and friendly when meeting new people and animals. Reward them for good behavior, like sitting nicely, instead of jumping around or being too excited.

2. **Establish basic obedience:** Before you introduce your puppy to new people or animals, make sure they know how to behave in basic ways. Teach them commands like "sit," "stay," and "leave it" so you can have more control over how they act around other people.

3. **Controlled introductions:** Keep things calm and in control when you introduce your puppy to new people or animals. Start by introducing your puppy to one person or animal at a time. As the puppy gets more comfortable, you can introduce them to more people or animals.

4. **Reward calm behavior:** When your puppy is calm during introductions, encourage the puppy and give it a treat. For example, if your puppy sits or stays calm when meeting a new person, you can reward it with treats, praise, or a gentle pat. This will teach your puppy that remaining relaxed leads to good things.

5. **Stop your puppy from jumping or getting too excited:** If your puppy tends to jump or get too excited when greeting people, stop this behavior. Don't pay attention to or reward

your puppy when it jumps. Instead, tell them to sit down or suggest something else they could do, like shake hands or give a high-five. Reward and compliment them when they do the right thing.

6. **Consistency and repetition:** Consistency and repetition are crucial to teaching your puppy to be friendly and calm. Practice making introductions often, with people and animals you know and with ones you don't. Repetition helps reinforce your desired behavior and gives your puppy more confidence in social situations.

7. **Controlled socialization:** Gradually introduce your puppy to different people and animals in safe places. This can be done through supervised play dates with well-behaved, vaccinated dogs or by introducing the dog to other people, kids and adults. Always watch the interactions and step in if anyone seems scared, angry, or uncomfortable.

8. **Seek out positive experiences:** Try to make good memories with new people and animals by ensuring your interactions with them are enjoyable and helpful. When people are calm and friendly with your puppy, ask them to pet it gently, give it treats, and provide it with praise. These good things help them to trust people and continue to do the right things.

9. **Controlled introductions:** Keep things calm and in control when introducing your puppy to new people or animals. Let them come up to you at their own pace, and watch their body language for signs that they are stressed or uncomfortable.

Socializing with people and animals of different ages: Introduce your puppy to people and animals of different ages, such as children, adults, and the elderly. This helps them learn how to act with people of different ages in a good way.

1. When introducing your puppy to children, it's important to prioritize their safety first and keep a close eye on everything. Children should learn how to treat the puppy with kindness and respect. Encourage kids to walk with the puppy calmly and not make any sudden movements or loud noises that might scare it. Teach them to hold out their hand for the puppy to sniff and allow the puppy to make the first move toward them. Reward both the puppy and the child for being calm and kind when they are together.

2. Let your puppy meet different adults, like family members, friends, and strangers. Start with people your puppy already knows and gradually bring in new people. Let adults come up to your puppy without being scared, and give them space to sniff and get used to them. Remind adults to speak softly. Reward your puppy for being calm and getting along well with adults.

3. Encourage calm and gentle interactions and ensure the puppy and the older person feel safe and comfortable during the introduction. Reward your puppy when he or she acts nicely and politely.

4. When introducing your puppy to people of different ages, do it in a controlled setting that encourages good interactions. This could happen at your house, in a neutral place, or wherever they feel most at ease. Ensure the puppy has enough room to come closer or move away as needed. Always monitor the interactions to ensure the puppy and the people involved are safe and healthy.

5. When your puppy meets people of different ages, try to give it good experiences. During these interactions, reward good behavior with gentle petting, treats, and praise. It's important to help your puppy associate different age groups with good things and teach them how to act in each situation.

6. Slowly introduce your puppy to people of different ages. Start with shorter, more controlled interactions, gradually making them longer and more complicated. This slow introduction helps your puppy gain confidence and learn to get along with people of all ages.

Understanding and Responding to Body Language:

To communicate with your puppy, you need to be able to read its body language. It lets you know how they feel, what they need, and how comfortable they are. A wagging tail can mean excitement or happiness, while a tucked tail could mean fear or worry.

When a dog's tail wags, it's usually because it's happy or excited. When a dog wags its tail, it is usually showing that it is happy. Depending on the dog and the situation, the way and how fast the tail wags can be different. A fast, wide wag usually means the dog is very excited, while a slower, more controlled wag can mean the dog is calmer or happier.

A dog's tucked tail is usually a sign of fear, worry, or anxiety. When a dog instinctively tucks its tail between its back legs, it is often a sign that it is giving up or feeling uncomfortable. There may be other signs of fear, like crouching, avoiding eye contact, or trying to hide, along with the tucked tail. Approaching a dog with a tucked tail carefully and giving them a calm, reassuring place to be will help them feel safe.

It's important to think about more than just whether the tail is wagging or tucked but also where it is and how it moves in different situations. For instance:

A stiffly raised tail may mean the animal is alert, sure of itself, or ready to fight. This is often what dogs do when they are on guard or trying to show who is boss.

A calm and happy state is often shown by a tail that hangs low and gently wags from side to side.

A high-held tail that moves quickly can signify excitement or anticipation. But it's essential to look at the dog's whole-body language and the way its tail moves to figure out if the excitement is good or if the dog may be too excited.

It's important to remember that a dog's tail position and wagging are not the only ways to tell how he or she feels. Dogs talk to each other by moving their bodies, making sounds, making faces, and wagging their tails. To understand how your dog feels, you need to know how they act and what is happening around them.

Ears: When ears are up, it means that puppies are paying attention. When ears are down, it means they are afraid or giving up. Please pay attention to how your puppy's ears are positioned to determine their feelings.

1. **Ears are up and facing forward:** When a dog's ears are up and facing forward, it usually means that it is paying attention and is on guard. This ear position shows that the dog is interested in what is going on around it and is paying close attention to what is going on. It usually means that they are paying attention, are curious, and are interested in their surroundings.

2. **Ears flattened or pinned back:** When a dog's ears are flattened against its head or pinned back, it usually means it is afraid, anxious, or giving up. This is an automatic response to situations that seem dangerous or stressful. When a dog's ears are pressed tightly against its head, it may be because it is afraid, worried, or trying to show that it is weak.

3. **Partially raised or semi-erect ears:** Some dog breeds have ears that aren't completely flattened or standing up. Even in these situations, the position of the dog's ears can still tell you a lot about how the dog is feeling. Partially raised or semi-erect ears can be a sign of alertness or mild interest, but it's important to look at other body language cues and the dog's overall behavior to get a good idea of how it feels.

4. **Relaxed and natural position:** When a dog's ears are in their natural position, which is neither too high nor too low, it usually means that the dog is happy and at ease. When a dog's ears are in a neutral position, it usually means that they are calm, happy, and not feeling any strong emotions at the moment. It's important to remember that the neutral position can be different for different breeds and types of ears.

5. **Context and other body language clues:** When trying to figure out what a dog's ear position means, it's important to consider the whole situation and other body language clues. You can get a better idea of how they feel by looking at their faces and tails and hearing what they say. For example, a dog with flattened ears, a cowering body posture, and avoidance behavior may be scared, while a dog with upturned ears, a wagging tail, and calm body language may be alert and interested.

Posture: A loose, relaxed body shows they are comfortable and sure of themselves. Stiffness or a slouched position could be signs of fear or stress. Respect your puppy's limits and give it space if it seems uncomfortable.

1. **Loose and relaxed body:** A dog with a body that is loose and relaxed is usually comfortable, at ease, and sure of itself. A loose body at ease may have a soft stance, a slightly bent or wagging tail, and a natural weight distribution on all four legs. When a dog's body is loose and relaxed, it shows that they feel safe and confident in their surroundings.

2. **The face that looks relaxed:** When a dog is calm and happy, its facial muscles tend to relax, and its eyes look soft. The dog's mouth might be slightly open, and the corners of its lips might be turned up in a gentle smile. Relaxed facial features show that a puppy is happy and at peace. This is usually accompanied by a flexible body and a tail that wags.

3. **Stiff or tense body:** On the other hand, a stiff or tense body can be a sign of fear or stress. A dog's body may look prim with tense muscles that give it a frozen or statue-like look. A stiff body is often a sign of fear, along with things like flattened ears, bigger pupils, and a tucked tail. This dog's body language shows that it is on guard and may feel a lot of anxiety or pain.

4. **Slouched or lowered body:** Slouching or lowering the body can also show that someone is afraid or giving up. When a dog's body is hunched, its head is down, its back is rounded, and they are in a cowering position, it usually means they are feeling scared or weak. This can be shown by flattened ears, a tucked tail, and avoiding direct eye contact. To help a dog who is in this position feel safe and secure, you should approach it with care and respect.

5. **Giving space:** It is very important for your puppy's health to respect their limits and provide them with space. If your puppy's body language shows signs of discomfort, fear, or stress, it's essential to give them a chance to go away or take a break. When you push a dog past its comfort zone, you can make its fear or stress worse, leading to bad experiences or even aggression. The best way to build your puppy's confidence is to give them a safe place to feel safe and let them try new things at their own pace.

Face language: Pay attention to how your puppy's face moves. When someone is calm and happy, their facial muscles are relaxed, and their eyes are soft. A dog may be scared or angry when it growls, snarls, or shows its teeth. Keep these signs in mind and change your approach accordingly.

1. **Fear, anger, or discomfort:** Growling, spitting, or showing teeth are often signs of fear, anger, or discomfort. When a dog growls, snarls, or shows its teeth, it means it feels scared or threatened. When a dog feels cornered, nervous, or protective, it may act in these ways. It's essential to stay out of the dog's way and not do anything to worsen the situation. In these situations, it's essential to give the dog space and let them calm down.

2. **Frown or lines on the forehead:** When a dog's forehead is creased, it could mean they are uncertain, worried, or trying to focus. When a dog is trying to figure something out or isn't sure about something in their environment, it might make this face. It's important to carefully approach a dog with a furrowed brow and reassure them to help them feel better.

3. **Wide eyes or dilated pupils:** Puppies often have wide eyes or dilated pupils when they are very excited or scared. When a dog's eyes or pupils get bigger, they are alert and may feel afraid or threatened. These visual cues show that you should be careful and approach the puppy gently to avoid causing more stress or anger. Dogs can feel safer if they have space and time to calm down.

4. **Eyes look smooth:** Eyes that are soft and relaxed are a good sign that the dog is calm, at ease, and comfortable. When a dog's eyes look smooth, it means he or she is feeling calm and happy. Relaxed body posture and other signs of relaxation, like a loose jaw and open mouth, often go with soft eyes.

Vocalizations: Dogs use sounds like barking, whining, and growling to express how they feel. For example, a whimper or whine may mean that the dog needs attention or is uncomfortable, while a lot of barking may mean that the dog is worried or on guard.

1. **Barking:** Barking is a natural way for puppies to talk to each other, and it can mean different things depending on the situation. It can mean different things, like letting someone know about a possible threat, showing excitement, trying to get someone's attention, or showing frustration. Different messages can also be sent by the barks' pitch, volume, and length. For example, a quick, sharp bark could mean that the dog is scared or trying to warn you, while a constant, repetitive bark could mean that the dog is bored or upset. It's important to pay attention to the dog's other actions and body language if you want to know what it means when it barks.

2. **Whining:** Whining is a high-pitched sound that dogs make to show how they feel and what they need. It can be a sign of discomfort, worry, fear, or a need to be noticed. For example, a dog may whimper or softly whine when it wants its owner to comfort it or pay attention to it. On the other hand, a long or loud whine could mean pain, discomfort, or that the dog needs to go outside to use the bathroom. To meet a dog's specific needs, it's important to look at the situation, watch the dog's body language, and act accordingly.

3. **Growling:** Growling is a sound that dogs make to show they are upset, warn others, or take a defensive stance. It's a way for them to tell each other that they feel threatened or upset. Growling can happen in different situations, like when a dog feels trapped, scared, or protective of food or toys. Growling is a clear sign that a dog is upset, so you should be careful when you approach one. Respect their limits and give them space to keep things from getting worse.

4. **Howling:** Some dog breeds are known for howling, which is a sound they make for different reasons. Dogs howl to talk to other dogs or to react to certain sounds, like sirens or other high-pitched noises. Dogs can also howl when they are scared, lonely, or trying to find other dogs to play with. It's important to notice that environmental factors can sometimes cause a dog to howl, so it's important to look at the dog's overall situation and behavior.

5. **Whimper:** Dogs can also whimper, yelp, or grumble. Sounds like whimpering or yelping are often signs of pain, discomfort, or worry. Grumbling or low rumbling sounds could be a sign that a dog is happy or a way for dogs to talk to each other, like when they are playing. Each of these sounds can mean something different depending on the situation, so it's important to look at the body language and behavior that goes along with it to get a fuller picture.

Body tension: A relaxed body shows someone is comfortable, while a tense or stiff body may show fear, anger, or discomfort. Respect their limits and make sure they are in a safe place.

1. **Respect personal space:** Puppies have their own space, just like people do. If your puppy shows signs of wanting space, like moving away, avoiding eye contact, or freezing, don't overwhelm or crowd them. Let them approach you.

2. **Signs of wanting space:** Puppies show their need for space in many subtle ways. Some common symptoms are moving away from people or other animals, avoiding direct eye contact, turning their head or body away, or standing still. These actions show that the puppy feels uncomfortable or too much and wants to get away. It's essential to pay attention to these signs and act correctly.

3. **Avoid overwhelming or crowding:** When a puppy shows signs of needing space, it's important not to overwhelm or crowd them. If you get too close to them or get in their way, it can make them feel stressed and lose their trust. Instead, let the puppy come to you on its own. Make a calm and welcoming space so they can feel more at ease and safe.

4. **Let them approach you:** Let the puppy decide when he or she is ready to come up and talk to you. You can make the puppy feel good and welcome by sitting or kneeling down, avoiding direct eye contact, and speaking in a soft, kind voice. Let the puppy come to you by sniffing or approaching you at their own speed. This makes them feel like they have more control over what's happening and boosts their confidence.

5. **Respect their boundaries:** It's important not to force the puppy to interact with you. If the puppy keeps moving away or avoiding contact, which is a sign that it wants space, it's essential to give it the space it needs. If you force them to do something before they are ready, it can make them feel more anxious or scared. Give them a place where they can go to relax and feel safe, and make sure they can get to their own space when they need to.

6. **Building trust and gradual socialization:** As time goes on and the puppy gets more comfortable and trusting, it may want to meet more people and socialize. Gradually introduce them to new things, people, and animals in a controlled and positive way. This lets them gain confidence at their own pace and start to think of social interactions in good form.

Positive reinforcement: When your puppy acts calm and happy, praise, treats, and gentle petting are all good ways to reinforce that behavior. This makes them more likely to keep sending out those good signs.

1. **Praise:** Verbal praise is a powerful way to acknowledge and reinforce your puppy's calm and happy behavior. Use a warm and excited tone of voice to show that you agree and are happy. For example, you can say in a cheerful voice, "Good boy/girl!" or "Well done!" When

you praise your puppy, it learns to link the praise with being calm and happy, which reinforces the behavior.

2. **Treats:** Treats are a tangible way to show your puppy how good he or she has been. As a positive reinforcement, give your puppy a small, tasty treat when calm and happy. Make sure the treat is something they really want, like a small piece of their favorite treat or a training treat made just for them. The treat is a quick reward, which makes them feel good about what they did and makes them more likely to do it again.

3. **Gentle petting:** Physical touch, like petting or stroking your puppy gently, can be a powerful way to reinforce calm and happy behavior. Dogs often feel safe and comfortable when they are touched by people they trust. When your puppy is relaxed and happy, you can softly stroke its fur or rub its belly. This kind of physical contact makes them feel even more comfortable and strengthens your relationship with them.

4. **Timing and consistency:** When your puppy act calm and happy, you should praise him and treat him gently immediately. This helps them make a clear connection between what they do and the good things that happen to them. Be consistent in your approach, and continually reinforce the behavior you want to see. Consistency and timely reinforcement help your puppy figure out what it did right and make them more likely to do it again.

5. **Repetition and practice:** It's important to practice often to strengthen the link between calm, happy behavior and positive reinforcement. Ensure your puppy has plenty of chances to be calm and happy throughout the day. For instance, praise them, give them treats, and pet them gently when playing or just chilling out. With practice, your puppy will learn that good things happen when he or she is calm and happy.

It's important to remember that each puppy is different, and their body language may change slightly from one to the next. Spending quality time with your puppy, watching how they act, and building a solid relationship with them will help you determine their unique cues and preferences.

Day 3: Leash Training and Walking Etiquette

Training the puppy to walk on a leash is an essential part of raising a well-behaved and obedient dog. Getting your puppy used to a collar and leash is the first step towards teaching them how to walk on a leash and keeping them safe when you're outdoors. When picking out the right equipment for your puppy, you should consider their size and breed to ensure they are comfortable and safe on walks.

Familiarize your puppy with the equipment:

Getting your puppy used to the gear is important for ensuring they are comfortable and willing to follow you on walks.

Sniff and check: Let your puppy sniff and check out the collar or harness first. Put it on the ground or hold it out so they can come up to it and look at it. This helps them get used to the new thing and calms any worries or fears they might have.

Gradual introduction: Once your puppy is comfortable sniffing and looking at the collar or harness, gradually introduce them to wearing it. Do these things:

1. Hold the collar or harness close to your puppy's neck or chest without fastening it. Give them treats and praise to show your appreciation for their help and good behavior.

2. Slowly and carefully put the collar or harness on your puppy while giving it treats and compliments. Take it off immediately after a short time, so they learn that wearing it is good.

3. Do this several times, gradually increasing the amount of time your puppy wears the collar or harness each time. Each time, give them treats and praise to help them keep thinking of the equipment in a positive way.

4. Once your puppy is used to wearing the collar or harness for longer periods, attach a leash and let them walk around inside. Give them treats and compliments.

Remember that getting your puppy used to the equipment takes time and a consistent process. Take it at their pace, always give them positive feedback, and make sure they enjoy the experience. This builds trust between you and your puppy and makes future walks more fun for both of you.

Proper fitting:

Fitting the collar or harness correctly and adjusting the length of the leash is essential for your puppy's comfort and safety. Here's a full explanation of how to make sure the right size:

Collar or Harness Fitting:

1. The collar or harness should fit your puppy's neck or chest snugly. It should be tight enough to keep them from slipping out of it but not so tight that it hurts or makes it hard for them to move. You should be able to fit two fingers between your puppy's neck and the collar or harness.

2. Measure the circumference of your puppy's neck with a soft measuring tape to make sure the collar fits right. Measure around where the collar would sit at the base of their neck. Refer to the manufacturer's size guide or measure your puppy to find the right size collar.

3. Measure the circumference of your puppy's chest just behind its front legs for harnesses. To choose the right harness size, use a soft measuring tape and the manufacturer's size guide. The legs and shoulders should be able to move freely in a well-fitted harness.

4. Most collars and harnesses have straps or buckles that can be moved to make the fit just right. Check to see if the collar or harness you choose can be adjusted, and make any necessary changes to ensure your puppy is safe and comfortable. Check the fit often and make changes as your puppy grows.

5. A collar or harness that fits your puppy well shouldn't hurt, chafe, or rub against its skin. Make sure there are no straps that hang down or are loose that could get caught on something when puppies walk. Check the collar or harness for signs of wear and tear often, and if it gets damaged, replace it.

Change the length of the leash:

Control and freedom: Changing the length of the leash is important if you want to keep control of your puppy while still giving it enough freedom to move around. It's a balance between letting them move around and making sure they stay close enough to you to be safe.

1. **Standard length:** A leash is usually between 4 and 6 feet long for walking. This length gives your puppy enough space to sniff, look around, and enjoy their walk while still letting you keep control of them and guide their movements.

2. **Environment and rules:** Consider where you'll be walking your puppy and what rules are in place. In busy streets or areas with a lot of people, it may be best to keep the leash shorter so you can keep a better hold on your puppy. Some places may also have rules about how long a leash can be, so find out what they are.

3. **Adjustable leashes:** Some leashes have features that let you change how long they are, like retractable leashes that let you change the length as needed. But be careful with retractable leashes because you may not have as much control over your puppy, and the thin cord can easily get tangled or break.

4. **Shorter leash:** The length of the leash should be based on your comfort and what your puppy needs. If your dog is small or you want to keep a closer eye on it, you might choose a shorter leash. A longer leash can give puppies that are bigger or more active more room to move around while still keeping them under control.

5. **Collar or harness:** Check out the fit of the collar or harness and adjust the length of the leash as needed. You can keep your puppy safe and comfortable on walks while still being in control and letting them enjoy their time outside by making sure the harness fits right.

Ongoing monitoring:

Puppies grow quickly, so it's important to check the collar or harness often to ensure it still fits. Measure their neck or chest every so often and make changes to the collar or harness as needed to keep it snug but comfortable. This way, the equipment won't get too tight or loose as your puppy grows.

1. **Acts and moves:** Pay attention to how your puppy acts and moves when you take him or her for a walk. If the puppy is pulling or scratching too much or trying to get out of the collar or harness, this could be a sign that it doesn't fit right. Fix any fit problems right away to ensure your puppy is comfortable and healthy.

2. **Skin and fur health:** Check where the collar or harness sits on your puppy's neck or chest. Look for signs like redness, irritation, or hair loss. If your puppy has skin problems, it may be because the collar or harness doesn't fit right or because you need to switch to a different one.

3. **Advice from a professional:** If you're unsure how to choose the right size or make it fit better, talk to a professional dog trainer or veterinarian. They can give you advice based on the breed, size, and individual needs of your puppy.

Handling and control of the leash:

When you walk your puppy, hold the leash firmly but don't pull too hard. Keep your grip loose but firm, allowing for some movement and flexibility.

1. **Training and guidance:** If you want your puppy to walk calmly on a leash, you need to teach it how to do so correctly. Use positive reinforcement techniques, like rewarding good behavior and discouraging pulling or too much tension on the leash. If you need to, get help from a professional to ensure that leash training is effective and safe.

2. **Think about the environment:** Change the length of the leash based on where you are. In places with a lot of people or traffic, keep the leash short so you can keep control of your puppy and keep it from getting close to potential dangers. You can lengthen the leash in open areas to let your puppy roam around more.

3. **Safety measures:** Always be aware of what's happening around you and the risks. Don't use a retractable leash near busy roads or other places. If your puppy tends to slip out of their collar or harness, you might want to use a second safety clip or a double-ended leash for extra security.

By checking how well the collar or harness fits and adjusting the length of the leash, you can ensure your puppy stays safe, comfortable, and under your control when you take him for walks. Ensure that your puppy's health and happiness come first and that their time outside is fun and rewarding for both of you.

Positive reinforcement for Handling and control of the leash:

Treats:

1. **Offer treats:** When your puppy wears a collar or harness, use small, tasty treats as rewards. You can give a treat right after putting on the equipment or at different times as it is being worn.

2. **When to give the treat:** After putting the collar or harness on your puppy, give it the treat. This makes a link between wearing the gear and getting the reward. Also, give your puppy treats every now and then while it's wearing it to reinforce the good feeling.

3. **High-value treats:** Use treats that your puppy likes and have a lot of value. These special treats can make wearing the collar or harness even more fun and help the puppy form a strong positive association with it.

Praise and petting:

1. **Verbal praise:** When your puppy wears the collar or harness, praise them happily and encouragingly. Say things like, "Good boy/girl!" or other encouraging things. The tone and words show how happy and pleased you are, which adds to the good experience.

2. **Pet your puppy gently:** Along with verbal praise, pet your puppy gently and show affection while it is wearing the collar or harness. Touch their head, back, or chin, whichever they like, to make a good feeling even stronger.

Play and fun activities:

1. **Play:** Play with your puppy while he or she is wearing the collar or harness. Play their favorite games, like fetch or tug-of-war, or do other things that are fun and interactive for them. This helps them connect the equipment with good times and reinforces their feelings about it.

2. **Fun walks:** While your puppy is wearing the collar or harness, take it on short walks or trips to look around. Let them sniff and explore their surroundings to keep their minds and bodies active. This ties the gear to fun outdoor activities, which makes the experience more positive and rewarding.

Consistency and repetition:

1. **Regular practice:** Consistency is the most important part of building a positive association. Practice putting the collar or harness on and taking it off often, and gradually increase the amount of time it is worn. The positive association will get stronger the more often your puppy gets positive feedback while wearing the equipment.

2. **Variety of positive experiences:** While your puppy is wearing the collar or harness, give it various beneficial activities to do. This can be done by going for walks in different places, playing with different toys, or doing exercises. The variety keeps things interesting and makes the excellent connection stronger.

Day 4: Basic Obedience Commands

Sit, Stay, and Lie Down:

It is essential for your puppy's safety, ability to communicate, and behavior as a whole to learn basic obedience commands. Here's a complete guide to teaching your puppy to sit, stay, and lie down:

Sit:

1. Choose a place that is quiet and free

2. Hold a small treat close to its nose to get your puppy's attention, and let it sniff it. This will make them curious. Make sure they know the treat is coming.

3. Hold the treat close to their nose and slowly move it up to their forehead. Their bottom will start to move toward the ground, and their head will tilt back naturally.

4. While your puppy is sitting down, say the word "sit" in a clear, firm voice. It's important to always use the same cue so that your puppy learns to link it to the action.

5. As soon as your puppy's bottom touches the ground and they are sitting, give them the treat and tell them, "Good sit!" or "Well done. Each time you train, do the process several times.

6. Be patient and give your puppy time to figure out what you want it to do after you tell it. With enough practice, they will start to think of the command as meaning "sit."

7. As your puppy gets better at sitting, reduce the number of treats you give as rewards. Instead, focus on giving affection, verbal praise, and petting as rewards. But keep giving treats every so often to reinforce good behavior.

8. Once your puppy has learned to sit inside, practice the command in different places with different distractions. This helps your puppy learn the behavior and ensures he can follow the command in different situations.

It takes time and patience to train. Don't scold or punish your puppy if it doesn't sit right away. Instead, you should stay calm and give them positive feedback. Your puppy will learn to understand and follow the "Sit" command if you consistently give positive feedback.

Stay:

To begin the training session, put your puppy in a sitting position. Use the "Sit" command, which was already explained, to make them sit down.

1. **Stay:** Say "Stay" and step backward with your palm facing them. With your puppy sitting, say "Stay" in a clear and firm voice. While saying the command, take a step backward and keep looking at your puppy. Your open palm facing them is a visual signal for them to stay still.

2. **Reward and praise for staying:** If your puppy stays in the same spot without moving, go back to them to offer them a treat and verbal praise. Rewarding them immediately following their stay reinforces the behavior you want and makes them more likely to stay in the future.

3. **Gradually increase the distance and length of the stay:** As your puppy gets more used to staying, gradually increase the distance between you and your puppy. Take a few more steps backward before returning to reward and praise them. Also, gradually increase the length of the stay by increasing the amount of time they need to stay in place before they get the reward.

4. **Always praise and reward:** It's important to keep reinforcing the stay command by going back to your puppy, giving them treats, and praising them for following the stay rule. This helps them understand that staying put is good and encourages them to keep doing what they are told.

5. **Corrections for unwanted movement:** If your puppy starts to move from the sitting position before you let them out of the stay, use a soft but firm "No" to tell them to return to where they were. Gently lead them back to the right spot, repeat the command, and start the stay exercise. Be patient and consistent when correcting unwanted movement, and always reinforce the behavior you want with rewards and praise.

Lie Down:

To begin training, place your puppy in a sitting position. You can tell them to sit down with the "Sit" command, which we already talked about.

1. **Hold a treat close to its nose and slowly lower it to the ground:** To get your puppy's attention, hold a small treat close to its nose. Slowly move the treat toward the ground, keeping it close to their body. This will make them want to follow the treat with their nose and start to get closer to the ground.

2. **Your puppy will naturally lie down as it follows the treat:** Keep lowering the treat toward the ground and let your puppy's nose follow it. As their nose follows the treat, their body will naturally lower towards the ground, and they will lie down.

3. **Say "Lie Down" and praise and treat them:** Say "Lie Down" in a clear, firm voice once your puppy is in the position of lying down. Right after you give them the command, give them the treat and tell them something nice, like "Good lie down!" or "Well done!" This helps them connect the command with the action and makes them more likely to do it again in the future.

4. **Practice the command often:** Tell your puppy to "lie down" over and over, and do it often. As they get better at following the command, you can give them treats less often. Instead of treats, give your puppy verbal praise and physical affection, like petting or stroking him gently. This helps them depend more on what you say and reinforces the behavior you want.

5. **Be patient and consistent:** If you want to teach your puppy to lie down, you need to be patient and consistent. Each puppy learns at its own pace, so give it time to figure out what you want it to do and do it right. Stay upbeat, use positive reinforcement, and don't punish or get angry with the puppy while you're training it.

Teaching Your Puppy to Come When Called:

1. **Start in a safe, quiet place:** Start the training in a safe, quiet place with few distractions. This lets your puppy focus on the training without getting too distracted by external stimuli.

2. **Get on the same level as your puppy and talk in a friendly way:** Crouching down or sitting on the floor puts you at the same level as your puppy. Say their name in a clear, friendly way to get their attention. Then, say "Come" in a friendly and inviting way.

3. **Use a reward to get them to come:** Hold out a treat or their favorite toy to get your puppy to come to you. Show them the reward and tell them you have something good for them.

4. **When they come to you, praise and reward them:** When your puppy starts to come toward you, praise their hard work and enthusiasm. Give them the treat or play with their toy as a reward when they get to you. This reinforces the connection between the word "Come" and getting a treat.

5. **Gradually add distractions and space between you and your puppy:** Once your puppy understands what it means to come when called in a controlled environment, gradually add space between you and your puppy. Start by moving back a step or two and practicing the command. Add mild distractions like low-level sounds or toys to make it harder for them to concentrate.

6. **Avoid yelling or punishing your puppy:** If your puppy doesn't come immediately or takes longer, don't yell or use negative reinforcement. This can make people scared and confused, which makes the training less effective. Instead, take your puppy back to a place with fewer distractions where he or she can succeed, and reward and praise the good behavior.

7. **Make sure that your dog's training sessions are short and fun:** Practice giving the "Come" command in different places, gradually making it harder for your puppy. Training works best when there is consistency, patience, and positive reinforcement. Your puppy will learn to come when you call it and have a strong recall response if you reinforce the command with treats and make a positive association with it.

Using Rewards and Positive Reinforcement:

When you train your puppy, rewards and positive reinforcement are very important. Here are some essential things to remember:

1. **Use treats with high value:** Choose treats that your puppy finds very appealing and fun. These treats should only be given during training to make training sessions more special and motivating. Small pieces of cooked chicken, cheese, or store-bought treats your puppy loves could be high-value. Your puppy will want to do what you want more if they really want the treat.

2. **Make a strong link:** You're reinforcing the link between good behavior and a good result by giving your puppy a treat right after they do what you want. This shows your puppy again that what they did gets them a treat, which makes them more likely to do it again.

3. **Timing is important:** Proper timing of treat delivery is important for successful training. Dogs have short attention spans, so rewarding them right after they do what you want them to do. If you don't give your puppy the treat at the right time, this can confuse the puppy and make it less likely to engage in good behavior.

4. **Use positive reinforcement:** One of the best ways to train is to use positive reinforcement because it rewards good behavior instead of punishing bad behavior. You can reinforce good behavior by giving your puppy a treat right after they do what you want them to do. This gives your puppy a good learning experience and encourages them to keep acting as you want them to.

5. **Change the treats:** Using treats as rewards work, but it's important to offer different kinds of treats during training. So, your puppy won't get bored or get used to just one kind of treat. Give them various treats with different tastes, textures, and sizes to keep them interested and motivated.

6. **Reduce your puppy's dependence on treats gradually:** As your puppy gets better at doing what you want him to do, you can give him treats less often. Switch from giving treats every now and then to giving verbal praise, physical affection, and other rewards that don't involve food. This shows your puppy that they will get positive reinforcement even when they don't get a treat.

Verbal praise and physical affection:

GOOD

When giving treats, say things like "Good job!" or "Well done!" in a cheerful tone. You can also reward good behavior by softly petting or rubbing their bellies.

You can use treats less and less as your puppy gets better at following commands. Instead, praise and affection should be used more and more.

Transitioning from treats to praise and affection: As your puppy gets better at following your commands, you can start to use praise and affection as rewards instead of treats. This is important because you don't want your puppy to be motivated only by treats. Instead, you want them to follow your commands because they want to please you.

1. **Reinforcing with praise:** Verbal praise is a very effective way to reinforce your puppy's good behavior. When they do what you ask them to do, give them verbal praise right away. For example, say "Good job!" or "Well done!" in a happy tone. This kind of praise helps strengthen the link between the behavior you want and the good that happens when you do it.

2. **Using physical affection:** Physical affection is another way to reinforce your puppy's good behavior, along with verbal praise. This can be done by gently petting the puppy, rubbing its belly, or patting it on the head. Your puppy can form a strong bond with you and feel rewarded when you touch them. This reinforces their good behavior.

3. **Rewarding with treats less often:** To move away from treats, gradually reduce the number of times you reward with treats and reward more with praise and affection. Start by giving your puppy treats every few times it does something you ask it to do correctly. Even if you don't always give your puppy a treat, this type of reinforcement keeps him or her interested and motivated.

4. **Maintaining treats as rewards:** Even if you're using more praise and affection, it's important to keep giving treats as rewards often. Randomly giving your puppy a treat for good behavior keeps him or her motivated and reinforces the idea that following your commands leads to good things. It keeps your puppy paying attention to your commands and adds an element of surprise.

Each puppy is different, and how they move from treats to praise and affection may depend on their personality and what drives them. Some puppies may respond well to more verbal praise, while others may be more interested in physical affection. Watch your puppy and figure out what works best for him or her, then change your approach to fitness.

Day 5: Chewing, Biting, and Digging Solutions

Understanding Chewing and Biting Behaviors:

Like babies, puppies learn about the world through their mouths. This is how they learn about their environment, things, and even other living things. They also use their mouths to ease the pain of teething and play with other puppies or people in their families. Understanding these deeper reasons for chewing and biting is important if you want to stop these behaviors.

Exploration: Puppies are naturally curious, and chewing gives them a chance to learn more about their surroundings. They put different things in their mouths to feel, taste, and figure out what they are made of. This helps them figure out what they can eat, what they can play with, and what they should stay away from. It is important to give them the right things to chew on and explore, like chew toys made for puppies.

1. When their adult teeth start to come in, puppies go through a phase called "teething," just like babies do. This can be a hard and sometimes painful process for them. Chewing helps them feel better and gives their sore gums some relief. Giving the puppies specific chew toys will assist you in getting them to concentrate on the proper objects to chew on.

2. Puppies chew and bite as part of normal play and social interaction. They explore their surroundings and play with their littermates or people who live with them. This helps them get close to each other, set up social orders, and learn not to bite too much. Encourage them to play in the right way and give them interactive toys to keep them from chewing or biting things that are not good for them.

3. But it's important to redirect these behaviors to good things and stop them from chewing or biting things that aren't good for them.

4. Give your puppies a variety of chew toys made for them helps them satisfy their need to chew. These toys should be safe, last a long time, and keep your puppy interested. You can keep them interested and happy by giving them different textures and shapes to chew.

5. It's important to praise and reward your puppy when it chooses to chew on the right toy or stops chewing on things it shouldn't. Verbal praise, affection, and treats every once in a while, can reinforce their good behavior and make it more likely that they will do it again.

6. Stop your puppy from chewing on things he shouldn't. If he starts chewing on things, he shouldn't redirect him. Give them a good chew toy to play with instead of the forbidden object, and try to get their attention off of it. Don't scold or punish your puppy because it could make it scared or confused.

7. Keep a close eye on your puppy, especially when he or she is exploring. Use baby gates or crate training to keep them out of rooms where they might see things they shouldn't. This lets you step in and change their behavior if you need to.

8. You can help your puppy learn what is okay to chew on and what it is not by understanding why it chews and bites and then using the right techniques to redirect it. With consistent guidance, positive reinforcement, and patience, you can shape their behavior and teach them good habits that will make your home a safe and happy place for both you and your puppy.

Teething: When a puppy's adult teeth come in, usually between the ages of 3 and 6 months, it hurts. This pain can be eased by chewing, so it's essential to give them the right toys to chew on during this stage.

1. Teething is a natural process that puppies go through as they transition from their baby teeth to their adult teeth. It usually happens between 3 and 6 months, but it can happen at different times for different puppies. During this period, the puppy's adult teeth begin to displace the baby teeth.

2. Puppies can have a hard time and sometimes even feel pain when they are teething. As their new teeth come in, their gums may get sore and sensitive, just like when babies get new teeth. This pain can make puppies chew more as they try to stop hurting themselves.

3. Chewing is a natural way to help puppies who are teething. It helps ease the pain by massaging their gums and giving them a good way to release the pressure caused by their teeth coming in. Puppies can help loosen the baby teeth and speed up the process of shedding them to make space for permanent teeth by chewing on the proper objects.

4. It's important to give your puppy the right toys to play with during this time. These toys should be made for teething puppies and should be safe and fun for them to chew on.

5. Look for toys with different textures that can give your puppy's gums different amounts of pressure and relief. Some toys have rough surfaces or ridges that can massage the gums, while others may be made of soft or flexible materials that are easy on sensitive areas.

6. Choose toys with mouths that are the right size for your puppy. They should be easy for the puppy to hold and move around, so it can chew on them comfortably. Avoid toys that are too small and could choke your puppy or are too big and hard for your puppy to play with.

7. Choose toys that will last and can stand up to the puppy's chewing. Look for toys made of strong materials like rubber or nylon that can stand up to the pressure that their growing teeth will put on them. Don't give your puppy toys that can be easily torn or shredded because they might eat small pieces.

8. Make sure the teething toys you choose are safe and don't have any chemicals that could hurt your baby. Check for small parts that could be swallowed or things that could cause choking. Check the toys often for signs of wear and tear, and replace them if they get broken or worn out.

9. When you first give your puppy a teething toy, use praise and positive reinforcement to get them interested. Show them how to chew on the toys by holding them in your hand or showing them the right way to chew. Watch your puppy while it plays to make sure it is using the toys correctly and not chewing on things that will hurt them.

It's important to remember that while teething toys can help, they might not stop the puppy from chewing completely. Puppies can still play with and chew on other things they can reach. To keep your puppy from chewing on things it shouldn't, puppy-proof your home by getting rid of potential dangers and giving it plenty of safe things to chew on. By knowing how teething works and giving your puppy the right toys, you can help him or her get through this stage with comfort and ease. Remember to be patient and consistent as you teach your puppy to chew on the right things. Soon, when they get their adult teeth, they will have learned how to chew in a healthy way.

Puppies use their mouths to explore their environment and play. It's how they find out about new textures, tastes, and things. It's important to show them what they can chew on and what they can't.

Supervise and redirect: If your puppy starts chewing on something it shouldn't, move its attention to a good chew toy in a calm way. Use positive reinforcement by giving them praise and treats when they chew on the right thing. You need to be consistent to teach them what they can chew on.

1. **Be patient and consistent:** Remember that puppies are still learning, and it takes them time to figure out what is okay to chew on and what is not. Keep changing their behavior, and don't scold or punish them because that can make them scared and confused. The best results will come from consistent training, using positive reinforcement, and working with patience.

2. **Use deterrents:** Sometimes, puppies will keep chewing on the same things even when they are told not to. In such situations, you can use taste deterrents available at pet stores. The bad taste of these substances makes puppies less likely to chew on things that have been treated with them. Always follow the directions; if needed, talk to a veterinarian.

3. **Provide both mental and physical stimulation:** Puppies need both mental and physical activity to keep them interested and keep them from getting bored. Make sure your puppy gets to play, go for walks, and play with toys that challenge their minds regularly. Do things that make you think about how to solve problems, like puzzle games or toys that give out treats. Mental stimulation can make them tired and take their minds off chewing or biting things that are bad for them.

4. **Establish a consistent routine:** Puppies do best with structure and performance. Set up a daily schedule that includes playing, training, and exercise times. A regular schedule keeps your puppy from getting bored and gives it a sense of security and purpose.

5. **Rotate the toys:** Make sure your puppy's toys are different and fun. Rotate the toys you already have so that they stay interesting. This keeps toys from getting boring for your puppy and keeps it interested. Interactive toys, like puzzle feeders or toys that give out treats, can keep your puppy busy and stimulate his mind.

6. **Use positive reinforcement:** When your puppy does something good, praise him, give him treats, or let him play with you. When they play without biting, praise them and give them something good to do. This shows that you get rewards and people's attention when you do good things.

7. **Ignore behaviors that your puppy does to get your attention:** If your puppy chews or bites to get your attention, it's important not to reinforce this behavior accidentally. Avoid giving your puppy attention or acting strongly when it acts this way. Instead, try to get them interested in a good toy or activity. Once they start to play with the thing, you want them to, praise them and pay attention to them.

8. **Make their environment more interesting:** Give your puppy a stimulating environment with places to explore and things to do. Set up puzzle feeders, interactive toys, or hide

treats around the house to keep your pet's mind active. Consider giving your puppy chew toys and mats with different textures and surfaces to satisfy its natural desire to chew.

9. **Consider professional help:** If your puppy keeps biting and chewing on things to get your attention or because it's bored, you might want to get help from a vet, professional dog trainer, or behaviorist. They can make plans and give you extra tools that are just right for your puppy and its needs.

Puppies may chew or bite to get attention or stop boredom. Ensuring they have enough mental and physical stimulation which can help them to stop from doing these things.

Redirecting Chewing and Biting:

Follow these steps to make sure your puppy bites and chews on appropriate things:

Offer the right chew toys: Give your puppy a variety of chew toys made just for them. Choose toys that will last, are safe, and will keep your puppy interested. Play with them together with these toys to get them to use them.

Getting the right chew toys is important if you want your puppy to chew on the right things. Here are some tips on how to choose and give your puppy chew toys:

1. **Choose toys that are safe and won't break:** Look for chew toys made just for puppies. Choose toys that are made of strong materials that your puppy can chew on. Don't give your child toys with small pieces that are easy to chew off and swallow because they could choke on them.

2. **Think about different textures:** When it comes to chewing, puppies like different textures. Give your puppy a variety of chew toys, like ones made of rubber, nylon, or rope, that have different textures. This lets your puppy try out new feelings and keeps them from getting bored with just one kind of toy.

3. **Toys that are the right size**: Make sure the chew toys are the right size for your puppy. Toys that are too small could cause your puppy to choke, and toys that are too big might be hard for your puppy to play with and chew on. Choose toys that are the proper size, for your puppy.

4. **Introduce the toys:** When introducing a new chew toy, play with it yourself to get your puppy interested. Show your puppy how excited you are about the toy and encourage it to play with it by holding it, shaking it gently, or throwing it. Puppies often act like their owners, so showing them how much they like the toy can encourage them to play with it and chew on it.

5. **Watch your puppy chew:** You should watch your puppy chew to make sure it is using the chew toy correctly. This gives you a chance to stop them if they start chewing on the wrong things or acting badly. Keeping an eye on how they chew also helps keep them safe and prevents accidents.

6. **Rotate the toys:** To keep your puppy interested, switch out their chew toys often. Put out some of the older toys while you bring out the new ones. This helps keep the toys interesting and keeps your puppy from getting bored with them. It also gives them a new and interesting way to chew.

7. **Clean and maintain toys:** Make sure your puppy's chew toys are safe and clean by checking on them often and cleaning them. Some toys may need to be washed or sanitized, while others may need to be replaced if they get dirty or broken. Your puppy will always have safe and fun things to chew on if you keep his toys in good shape.

Remember that you need to use positive reinforcement to get your puppy to chew on the right things. When your puppy chews on their own toys, praise and reward them with treats, words of encouragement, or time to play. This helps them learn which things are okay to chew on and reinforces the behavior.

Avoid punishment: When your puppy chews on things it shouldn't, it's important not to hit or scold it. Instead, turn their attention to a good toy and praise them when they behave well.
When your puppy chews on things they shouldn't, it's important not to hit or yell at them. Giving

punishment can backfire and cause your puppy to be afraid, anxious, or even aggressive.

Preventing and Managing Digging:

Digging is something that dogs do naturally, but it can be a problem if they do it in the wrong places, like your garden or carpet. Here's how to stop and deal with digging:

1. **Provide alternatives:** Make a place in your yard for your puppy to dig where it can satisfy its natural urge. Place treats or toys in this area to get their attention. Give them a treat when they dig in the right place.

2. **Designated digging area:** A good way to change your puppy's behavior is to give them other ways to satisfy their natural urge to dig. You can get them to dig in the right place by making a designated digging area in your yard.

3. **Choose a specific place to dig:** Choose a spot in your yard where your puppy can dig without bothering you. This area should be easy to get to and away from delicate plants or gardens. The soil should be soft and easy to dig in.

4. **Set up the area to be dug:** Clear the area of any trash, rocks, or other things that your puppy could hurt itself on while digging. Loosen the soil or add sand to make it look better and make digging easier.

5. **Make it look good:** To get your puppy to dig in the right spot, sprinkle treats on the ground or hide their favorite toys in the dirt. When they smell rewards and see them, they will be more interested in the area.

6. **Stop your puppy from digging in the wrong place:** To get your puppy to dig in the right place, gently guide them there if you see them digging somewhere else. Use a firm but gentle voice command like "Don't dig" or "Dig here" to lead them to the right place. Once they start digging where they should, praise them and give them a treat.

7. **Consistency and repetition:** Consistency and repetition are important when teaching your puppy to dig in the right place. When they dig in the right place, reward them. This will help them keep doing what you want them to do. With time and practice, they will start to prefer the designated area and learn that it is where they should dig.

Remember that puppies have a natural urge to dig, so it's important to give them a place to do it..

Exercise and mental stimulation: To keep your puppy from getting bored, make sure it gets enough physical exercise and mental stimulation. When a puppy is tired, it is less likely to do things that are bad for it, like digging. Make sure your puppy gets enough physical and mental exercise so it doesn't get bored and stops doing bad things like digging.

Puppies have a lot of energy and need regular physical activity to burn it off. They stay active, healthy, and mentally stable when they do physical activities. Think about these things:

1. **Daily walks:** Take your puppy on regular walks outside so he or she can learn about the world around them, meet other dogs and people, and use all of their senses.

2. **Playtime:** Use toys like balls, ropes, or interactive puzzles to play with your puppy in a way that involves both of you. Play fetch, tug-of-war, or hide-and-seek with them to keep them moving and their minds active.

3. **Dates for puppies:** Schedule playtime with other vaccinated puppies or dogs that are friendly. This gives your puppy an opportunity to play with other dogs and meet other individuals.

4. **Activities for dogs:** You might want to sign your puppy up for agility training, obedience classes, or fly ball. These things will keep your puppy's mind active, help you get closer to it, and challenge its physical abilities.

Mental stimulation: Along with physical exercise, your puppy needs mental stimulation to stay interested and avoid getting bored. Your puppy will be less likely to do something bad, like dig, after performing mental exercises. Here are some ways to keep their mind active:

1. **Puzzle toys:** Buy puzzle toys for your puppy that are interactive and require him or her to solve problems or find hidden treats. These toys keep their minds busy and teach them how to solve problems.
2. **Training sessions:** When you train your puppy regularly, you not only teach it basic commands, but also keep its mind active. Teach them new tricks or obedience commands, or do clicker training with them to keep their minds busy and on task.
3. **Food-dispensing toys:** Use toys that give out food or treat balls to make mealtimes more fun and challenging.
4. **Introduce your puppy to new places:** Introduce your puppy to new places, things, and experiences to get them interested in new things. Take them to different places, let them smell and touch new things, and slowly introduce them to different stimuli in a controlled and positive way.
5. **Consistency and routine**: Set up a daily routine that is the same every day and includes both physical exercise and mental stimulation. Set up regular times for walks, playtime, training, and other activities to give your puppy structure and predictability. Consistency helps them know what to expect, lowers their anxiety, and helps them live a well-balanced, happy life.

Remember that the amount of exercise a puppy needs depends on its breed, age, and energy level. It's important to make sure that your puppy's activities and stimulation meet his or her needs. By giving your puppy regular physical exercise and mental stimulation, you can keep it healthy and less likely to do things like digging out of boredom.

Day 6: Advanced Training Techniques.

Teaching More Complex Commands:

Basic obedience training gives your puppy the skills he needs to learn more complicated commands. Here's how to teach and introduce more complicated commands:

1. **Target Training:** Teaching your puppy to touch or follow a specific target, like your hand or an object, is called "target training." This method can be used to teach more complicated tricks and commands. Start by putting the target in front of your puppy and giving it a treat when it touches or follows it. Step by step, make it harder by moving the target to different heights or places. This technique helps people pay attention, move together, and respond better.

2. **Clicker Training:** Clicker training is a popular method that uses a clicker, which is a small device that makes a clear clicking sound, to show when a behavior is good. By linking the clicker's sound to a treat, you can tell your puppy exactly when he does what you want him to do. Clicker training is a great way to change behaviors that are hard to change, and catch exact moments of success. It makes communication clear and speeds up the learning process.

3. **Backward Chaining:** This is a way to teach complex tasks or steps by going from the last step to the first one. Make the behavior easier to do by breaking it down into smaller steps. Slowly add the steps before the last one in the opposite order, starting with the last one. This method helps your puppy know what you want him to do and

reinforces each step along the way. Backward chaining is especially useful for tasks like getting things or doing tricks that are hard to explain.

4. **Duration and Distance Training:** Duration and distance training builds on basic commands to help your puppy keep a behavior going longer and does it from a farther distance. Start by giving your puppy short commands and rewarding it when it stays in position for longer periods. In the same way, work on getting farther away from your puppy while keeping their response to commands the same. These techniques help people pay attention, stay in control, and be reliable.

5. **Proofing:** To make sure your puppy's training works in real life, you need to expose it to different distractions, environments, and situations. Gradually add distractions to training sessions, such as noises, other animals, or new people. Practice commands in different places and conditions to make sure they can follow them everywhere. That way, your puppy will always learn to listen and do what you say.

6. **Boundary training:** This teaches your puppy to stay in a certain area or stay away from certain places. Start by selecting up physical markers or visual cues to show where the boundaries are. Use positive reinforcement to praise your puppy when it stays within its boundaries and redirect it when it tries to go outside of them. Boundary training is beneficial for making sure everyone is safe, keeping people out of certain areas, or setting up no-go zones.

7. **Task Discrimination:** Teaching your puppy to understand the difference between different tasks or commands is called task discrimination. For example, if you have taught your puppy to "fetch" and "sit," you can challenge to fetch an object first and then sit before they get a reward. This method improves their ability to think, makes it easier for them to understand different cues, and makes their minds more flexible.

8. **Hand signals and silent commands:** Instead of verbal commands, you can use hand signals and silent commands. Combine certain hand gestures or looks with commands your puppy already knows. Slowly stop giving verbal cues, but keep giving hand signals or visual cues. This method lets you talk without making noise, which is helpful in noisy places or for puppy's with hearing problems.

9. **Problem-Solving Games:** Play problem-solving games with your puppy that teach it to think for itself and make decisions. Hide treats or toys and tell your puppy to use its sense of smell or problem-solving skills to find them.

Splitting up complicated commands: Splitting up complicated commands into smaller steps is a good way to make sure your puppy learns them well. When you teach your puppy, a complicated command like "Roll over," it can be hard for them to understand the whole sequence at once. You can make it easier for them to understand and move forward if you break it down into smaller steps.

1. **Start with a simple order:** Start by going over the basic command, which is the basis for the more complicated command. In this case, you should make sure that your puppy knows the "Lie down" command. Practice this command in different places and situations to make sure your puppy knows how to follow it.

2. **Introduce the next step:** When your puppy is good at the "Lie down" command, you can move on to the "Rollover" command, which is the next step. Start by giving your puppy a treat to get it to lie down. Hold the treat close to the puppy 's nose and move it in a circle toward the puppy 's shoulder. This will help the puppy roll over onto its side.

3. **Add the verbal cue:** When your puppy is getting the hang of rolling over, say "Roll over" or whatever cue you like. Just before you move your puppy into the rolling position, say the cue. Repeat this step several times, always giving the same verbal cue.

4. **Reinforce with treats and praise:** Every time your puppy rolls over successfully, give it a treat and a lot of enthusiastic praise. Positive reinforcement helps your puppy keep doing what you want him to do and encourages him to keep learning. As your puppy gets better, you should use treats less and praise him or her more verbally.

5. **Practice and refine:** To practice the command, repeat the steps often. Be patient and give your puppy time to figure out what's going on and get used to it. As they get better, you can improve the command by relying less and less on treats and more and more on your voice.

6. **Be consistent and patient:** Remember that breaking up complicated commands into smaller steps lets your puppy learn at its own pace and gain confidence. During the training process, it is important to be consistent, patient, and upbeat. If your puppy is having trouble with a certain step, go back to the step before it and make sure they understand it.

With time, practice, and praise, your puppy will be able to learn and follow the complicated command "Roll over."

Use consistent cues: Use the cue whenever you give a command so your puppy learns to connect it with your desired behavior.

When it comes to advanced training techniques for puppies, there are a few ways to improve their skills and abilities even more. These methods go beyond simple commands and help teach more complicated behaviors. Here are some ways you can do that:

1. Shaping is a training method in which you shape the behavior you want by reinforcing small steps or close versions of the final behavior. For example, if you want to teach your puppy to bring you something, you can start by giving it a treat when it picks it up. Then, you can shape the behavior by asking it to get the object closer to you and then put it in your hand.
2. To target, you teach your puppy to touch a certain object with its nose or paw, like your hand or a target stick. This is an excellent way to teach more complicated commands, like how to turn on light switches or close doors. Start by showing your puppy the target object and giving it a treat when it touches it. As time goes on, your puppy will learn to do more complicated things.
3. Back chaining is a way to teach a behavior by starting with the last step and working your way backward. This is useful for complicated actions that happen in a particular order. For example, if you want your puppy to spin, sit, and lie down, you would teach them to lie down first, then sit, and then spin, making sure they got each step right before moving on to the next one.
4. When you proof a behavior, you practice it in different places, with distractions, and under different conditions to make sure your puppy can do it reliably, no matter what. Gradually add distractions or challenges while reinforcing the desired behavior. This will help your puppy learn how to behave in a variety of situations.
5. Task training is when you teach your puppy skills or tasks that can be useful or even save its life. This can include things like getting things, opening doors, letting someone know when they hear something, or helping them move around.

Remember that advanced training methods require patience, consistency, and positive reinforcement. Divide hard, challenging behaviors into smaller steps that can be done, and give clear cues or commands. Use treats, praise, and play to reinforce the behaviors you want, and always put your puppy's happiness and well-being first during training.

Impulse Control and Patience Training:

Your puppy needs to learn to control his or her impulses and be patient for self-control and overall obedience. Here's how to work on controlling your impulses and being patient:

1. **"Wait" command:** Teach your puppy the "Wait" command so that they wait before getting food, going through doors, or getting out of their crate. Start by telling them to sit or stay, then say "Wait," and use your hand to tell them to stay where they are. Before letting them move on, gradually lengthen the time they must wait.

2. **"Leave it" command:** Teach your puppy the "Leave it" command, which tells them to stop playing with something they are interested in. Start by putting something small on the ground and covering it with your hand. Tell them "Leave it" and treat them if they don't touch the object. Step by step, make it harder by using more tempting things and practicing in different places.

3. **"Stay" command:** Increase the distance and length of the basic "Stay" command. Start with short times and short distances. As your puppy gets more comfortable and reliable with the command, you can gradually increase the time and distance.

4. **Practice impulse control in everyday situations:** Use everyday situations to practice controlling your impulses. Ask your puppy to wait before going through a door or getting their food bowl. Rewarding them when they do what you ask helps them learn to be patient and self-controlled.

Recall Training:

Step 1: Choose a Quiet place.

- Start training recall in a calm, familiar place, like your home or a quiet backyard. Your puppy can focus on learning the recall command by cutting down on distractions.

Step 2: Hook up a light leash.

- Attach a light leash to the collar or harness of your puppy. If they need it during training, this leash will gently lead them in the right direction.

Step 3: Use High-Value Treats

- Keep plenty of valuable treats on hand. These treats should be especially tasty to your puppy and only be used for training him to come when called.

Step 4: Say the Recall Command

- Call your puppy's name and then say the recall command, such as "Buddy, come," while standing close by. Use a happy, encouraging tone of voice to make them want to talk back.

Step 5: Reward and encourage

- As you say the recall command, show your puppy that you want it to come to you by doing things like crouching down and opening your arms. When your puppy starts to move towards you, keep talking to them happily and excitedly.

Step 6: Give praise and rewards

- As soon as your puppy gets to you, please give them a treat they really like and lots of praise. Positive reinforcement makes the behavior you want to see more likely to happen again.

Step 7: Gradually Increase the Distance

- Once your puppy consistently comes when you say "recall" in a controlled environment, you can start to move farther away from it. Start by moving a few steps away from them, and as they get more reliable, move farther away.

Step 8: Introduce Mild Distractions.

- As soon as your puppy is good at coming when called from different distances, you can start to introduce mild distractions. Start with simple tasks, like throwing a nearby toy or making a soft noise. If your puppy gets distracted, use the leash to gently bring it back to you and remind it of the recall command.

Step 9: Phase out the Leash

- As your puppy gets better at coming when you call, you can slowly stop using the leash. Start by doing recall exercises in a safe place without the leash, but keep it nearby if you need to use it. At some point, your puppy will be able to respond reliably even when it's not on a leash.

Step 10: Maintain consistency and continue performing it.

- For recall training to work, you need to be consistent. Use the same recall command every time, and ensure everyone in your home uses the same recall command. Keep doing recall exercises regularly, reward and praise the behavior, and reinforce the recall command for as long as your puppy lives.

Remember that the keys to successful recall training are patience and positive reinforcement. Celebrate your puppy's progress and stay positive and helpful throughout the process. With consistent training and praise, your puppy will learn to come to you when you call, ensuring they are safe and giving you peace of mind.

Teach your puppy fun and challenging tricks like "fetch," "roll over," "shake hands," or "play dead.

Select a Toy: To begin, choose a toy or ball your puppy enjoys playing with. It could be a stuffed animal, a rubber ball, or any other secure and suitable toy.

Involve Your Puppy: Introduce the toy to your puppy and allow them to sniff and interact with it. You can make it more appealing to them by gently shaking or moving the toy. This will contribute to the toy's interest and excitement.

Fetch:

Introduce the Cue Word: Choose a cue word or phrase to associate with the fetching behavior, such as "fetch," "get it," or any other word you choose. This cue word will be used to instruct your puppy to retrieve the toy and return it to you. If you have one, you can also use it as a training tool.

Initial Toss: To begin, place the toy or ball in front of your puppy's nose and say the cue word clearly and confidently. This helps them associate the cue word with the fetching action. Toss it a short distance away (just a few feet away) in an area where they can easily reach it once they are focused on the toy.

Encouragement and Praise: When you toss the toy, use an excited and positive tone of voice to encourage your puppy to go after it. You can also guide them with gestures, like pointing to the toy. If your puppy begins to move toward the toy, praise and reinforce him by saying "Good job" or "Good fetch!" This positive reinforcement assists them in understanding that they are acting appropriately.

Encouragement and Praise: When your puppy reaches the toy, he or she may sniff or chew on it. Use the cue word "fetch" again cheerfully to teach them to bring it back to you. Squat down, open your arms, and offer a treat to incentivize them to return with the toy. When they approach you with the toy, give them verbal praise and a treat.

Repeat and Increase Distance:

1. Repeat the process, gradually extending the tossing distance.

2. Begin with short throws and gradually increase to longer throws as your puppy gains comfort and confidence in retrieving the toy.

3. Remember to use the cue word "fetch consistently" and to provide positive reinforcement when they bring the toy back to you.

Training takes time and patience, so practice the fetch exercise regularly. Keep training sessions brief and enjoyable, and end on a positive note. With consistent practice, your puppy will learn to associate the cue word with retrieving the toy and returning it to you.

Roll over:

Start with Your Puppy Lying Down: Make sure your puppy is calm and at ease by putting it on its belly. You can choose a place to train your puppy that is comfortable and quiet, so it can pay attention without being distracted.

Bring up the Treat: Hold a treat close to its nose to get your puppy's attention. Make sure it's a treat that your puppy really wants and will get excited about. This will be their reward for being able to do the roll-over behavior well.

Lure Towards the Shoulder: Move the treat slowly from your puppy's nose to its shoulder. The goal is to get them to flip over. As their nose follows the treat and they start to shift their weight, their body will start to roll over on its own.

Cue Word and Reward: When your puppy starts to roll onto its side, use a cue word like "rollover" or any other phrase you like. Be consistent with the word you choose as a cue so your puppy can learn to connect it with the action. At the same time, praise and encourage them in a happy and excited voice. Give them the treat as a reward after they roll onto their side. This will help them keep doing what you want them to do.

Move the Treat Gradually: With each repetition, move the treat a little closer to your puppy's back. The goal is to get them to roll all the way over onto their backs. Make sure always to use the cue word during the process.

Practice and reinforcement: Do the same thing repeatedly, slowly moving the treat closer to your puppy's back each time. Every time they roll over, tell them how great they did, and give them a treat as a reward. Positive reinforcement is the best way to teach your puppy that rolling over is good.

Step-by-Step Progression: If your puppy has trouble rolling over, you can teach the behavior in small steps. At first, focus on rewarding partial rolls to the side, and work your way up to a full roll onto the back. With time and practice, your puppy will be able to do a complete rollover with more ease and confidence.

Patience and consistency: Training takes time and practice, so do the roll-over exercise often. Short, positive training sessions will keep your puppy interested and motivated. Each puppy learns at a different pace, so be patient with your puppy.

Remember that it's essential to make training a positive and encouraging place. Give your puppy a treat when they do something good, and always end each session on a good note. Your puppy will eventually learn to roll over on command if you use and treat the cue word often.

Shake hands:

Start with Sit Position: Have your puppy sit in front of you calmly and attentively. Make sure they can pay attention and are ready for the training. You can use treats to get them to sit down if needed.

Prepare the Treat: Close your hand into a fist and hold a small treat. Ensure your puppy can smell the treat, but they can't get to it yet. This will get their attention and encourage them to participate in the training.

Present Your Closed Hand: Hold your closed hand toward your puppy's paw. This will make them look at your hand to see if there's a treat inside.

Cue Word and Paw Interaction: When your puppy sniffs or paws at your closed hand, say a cue word like "shake" or "paw" clearly and consistently. This will help them connect the cue word to the behavior you want them to do. Just be patient and wait until they touch your hand.

Open Your Hand and Reward: As soon as your puppy touches your hand, open your hand and give them the treat as a reward. You can also use verbal praise and positive reinforcement, like saying "good job" or "good shake," to maintain your desired behavior.

Repeat and reinforce: Do the process more than once, each time using the cue word, putting out your hand, and giving your puppy a treat when they touch it. This helps them learn that pawing at your hand is a good thing to do that will get them a treat.

Gradual Reduction of the Treat Prompt: Once your puppy consistently gives you their paw when you hold your closed hand, you can start relying less on the prompt. Start by putting your hand out sometimes without a treat in it. If your puppy still gives you their paw, praise them verbally and give them treats from other places sometimes.

Practice and generalization: Teach your puppy the "shake hands" command in different places and with other people so they can do it in any situation. This helps them understand that when they are told to shake hands, they should do so no matter what.

Patience and consistency: Training takes time and patience, so use the same cue word and reward system every time. Keep your puppy's training sessions short, positive, and fun. Always make sure to end the session on a good note.

By doing these steps and always rewarding your desired behavior, your puppy will learn to shake hands when you tell it to. Over time, they will know that the cue word means to offer their paw and will get praise and treats for a good "shake hands" response.

Play dead:

Start with Your Puppy Lying Down: Start with your puppy lying in a calm and relaxed position. Choose a place to train your puppy that is quiet, comfortable, and free of distractions.

Introduce the Treat: Put a treat close to your puppy's nose to get their attention. Use a treat that will get them excited and interested. This treat will be a reward for doing the "play dead" behavior correctly.

Tilt the Head: Move the treat slowly from your puppy's nose to the side of its head. The goal is to get them to tilt their heads to watch the treat move. As their head tilts, their body will naturally move a little bit.

Cue Word and Reward: When your puppy tilts its head, use a cue word like "play dead" or any other phrase you like. Be consistent with the word you choose as a cue so your puppy can learn to connect it with the action. At the same time, praise and encourage them in a happy and excited voice. Once they tilt their head, treat them to reinforce the behavior you want to see more of.

Gentle Roll onto the Side: Roll it gently onto its side while your puppy is still lying down. Gentle support and making sure they feel safe and comfortable. This step will simulate the "playing dead" position. While they are on their side, keep praising and giving them treats and verbal praise.

Repeat and reinforce: Do the process repeatedly, combining the head tilt and rolling to the side more and more each time. When your puppy successfully tilts its head and lies on its side, use the cue word, give it a treat, and praise it verbally. This helps them connect the cue word with the behavior they should be doing and reinforces the order of actions.

Practice and improvement: Your puppy will get used to the whole "play dead" sequence as he or she does it more often. Gradually lengthen the time they have to stay in the "playing dead" position before you give them something. You can also use a visual cue, like pointing your finger like a gun or saying "bang," to make the behavior more fun.

Patience and consistency: Training takes time and effort, so do the "play dead" exercise often. Training sessions should be short and upbeat and always end positively. Each dog learns at a different pace, so be patient with your puppy.

During training, remember to create a positive and encouraging environment. Treat your puppy when they do something right, and make training fun. With consistent practice and positive reinforcement, your puppy will learn to connect the cue word with the "play dead" behavior and act accordingly.

Teaching Your Puppy to Stay Calm during Meal Times:

Practice Waiting for Food:

1. Before giving your puppy its meal, tell it to "sit" or "wait" calmly.

2. Hold the food bowl or treat in your hand and wait until your puppy is still and calm.

3. Set down the food bowl and treat your puppy only when it is sitting still or waiting patiently.

Increase the waiting time over time: At first, start with short waiting times, like a few seconds, and slowly make them longer. Use a timer or mentally count to make sure you stay on track. Reward your puppy for being calm and patient while they are waiting.

Address Excitement and Jumping: If your puppy gets too excited or jumps up during meal time, take the food bowl or treat away calmly and wait for them to calm down before trying again. This makes it clear that you must be calm to get the food.

Encourage Gentle Eating: Show your puppy how to eat slowly and calmly. If they start eating too fast, you can slow them down by giving them interactive toys or puzzle feeders. The puppy has to work for its food with these toys, which is good for its mind and makes mealtimes calmer.

Practice the "Leave it" command:

1. Teach the puppy the "leave it" command to stop it from grabbing or begging for food while you eat.

2. Start by closing your hand around a treat and saying, "Leave it."

3. When your puppy stops trying to get the treat, give them a different treat from your other hand to show them how much you appreciate them.

4. Gradually work up to practicing the command with plates or tables full of food.

Reinforce Positive Behavior: Always treat your puppy for being calm and patient during meal times. Use treats, words of praise, and gentle petting to remind them of how well they are doing. This helps them connect being calm with rewards, which makes them more likely to keep doing what is wanted.

Stay Calm and Patient: You should stay calm and patient during training. Don't scold or punish your puppy if it gets too excited or acts up while eating. Instead, try to change their behavior and encourage them to stay calm.

Adjust training techniques: If your puppy has trouble with a certain command, try breaking it down into smaller steps or changing how you train. Try out different things to see which ones work best with your puppy.

Talk to a professional dog trainer or behaviorist: This is especially important if you're having trouble with your puppy for a long time or if its behavior becomes scary or hard to control. They can give you expert advice, personalized training plans, and solutions that are just right for your puppy.

Problem-solving and troubleshooting:

Managing common behavior problems during advanced training:

Taking your puppy's training to a higher level can present challenges. As you work to improve their skills and teach them more complicated commands, they may start to act up. In this section, we'll talk about common behavior problems that come up during advanced training, as well as good ways to deal with them.

1. **Problem:** Your puppy may have trouble staying on task during advanced training sessions, especially in environments that are distracting or when he or she is exposed to new things.

Solution:

- **Gradual Exposure:** Gradually expose your puppy to distractions by starting in a controlled, low-distraction environment and gradually making it harder. This teaches them how to maintain focus despite the presence of distractions.

- **Make it more interesting:** Use treats or toys with much value to keep your puppy interested and on task while training. Give them a lot of praise and rewards for paying attention and ignoring distractions.

- **Marker Training:** Use clicker training or verbal markers to encourage focused behavior. Reward your puppy as soon as they respond to the marker. This will help them pay attention and stay on task.

2. **Difficulty Generalizing Commands:** Problems with following commands in different places or with different people: Your puppy may find it hard to follow commands in different places or with different people, making it hard to generalize their training.

Solution:

- **Practice in different places:** Gradually teach your puppy commands in different places, like parks, streets, or busy areas, to help them learn to follow them anywhere. Start with commands you already know and slowly make them harder.

- **Practice with different handlers:** Have different family members or friends practice giving your puppy commands. This helps them learn how to respond to different people's commands, making it easier for them to use their training in different situations.

- **Use rewards that change:** When your puppy does what you ask, give it different rewards. This keeps them from getting used to certain rewards and helps them better understand the command.

3. **Reinforcing Behaviors You Don't Want:** During advanced training, making bad habits stronger by accident can slow down progress. Your puppy might figure out that specific actions get him attention or treats, even if those actions are bad.

Solution:

1. **Figure out what reinforces bad behavior:** Look at how you interact with the person to figure out what unintentionally reinforces terrible behavior. For instance, if the puppy jumps on you to get your attention, don't ignore it until it's calm.

2. **Ignore Unwanted Behaviors:** If your puppy does something you don't like, like barking or jumping too much, don't pay attention to them or give them treats until they do what you want. Once they calm down or do what you want them to do, praise and reward them.

3. **Redirect and Replace:** Teach your puppy other things to do that aren't the same as bad behavior. For example, if they like to chew on furniture, give them appropriate chew toys to play with instead and give them treats when they do.

4. Regression in Training: Problem: Your puppy may sometimes forget commands or behaviors it has already learned. This is called regression.

Solution:

1. **Reinforce Basic Skills:** Go back to the basics and help your puppy remember the basic skills he or she has already learned. This helps them feel better about themselves and understand commands better.
2. **Gradual Progression:** Once your puppy is good at the basics again, slowly start teaching it more advanced commands. Don't give them too much at once, and build up their skills gradually.
3. **Patience and consistency:** When you train, be patient and consistent. Regression is standard and can be caused by distractions or changing routines. Your puppy will get back on track if you are consistent and give him or her positive feedback.

5. How to deal with fear or anxiety: Some puppies may show fear or anxiety during advanced training, slowing them down and hurting their overall health.

Solution:

1. **Gradual desensitization:** If your puppy shows fear or anxiety around certain objects, sounds, or situations, gradually expose them to these triggers in a controlled and positive way. Start far away or with a less intense version of the trigger. Reward calm behavior and slowly move closer or make the trigger stronger.

2. **Counterconditioning:** Pair what your puppy is afraid of with something it likes, like treats or playing. This helps make good memories and, over time, lessens fear or anxiety.

3. **Get Professional Help:** If your puppy's fear or anxiety is severe or doesn't go away, talk to a professional dog trainer or veterinary behaviorist. They can give you advice and techniques that are tailored to your puppy's needs.

6. Overexcitement or Impulsivity: Some puppies may have trouble being too excited or too impulsive during advanced training, making it hard to keep control and stay on task.

Solution:

1. **Make the rules clear:** Set clear rules and limits for your puppy to know what is expected of them. Give them clear cues or commands when they need to calm down or settle down.

2. **Impulse Control Exercises:** Do exercises like "sit-stay" or "leave it" to help your puppy learn to control their impulses and improve their ability to pay attention and follow commands.

3. **Mental Stimulation:** Give your puppy mental stimulation with interactive toys, puzzle games, or scent work to help them use up their extra energy and stay focused during training.

7. Problem: During advanced training, it can be hard to handle leash reactivity, which is when your puppy gets angry or aggressive toward other dogs or people while on a leash.

Solution:

1. **Positive reinforcement:** Use positive reinforcement to praise your puppy when he or she is calm and doesn't act up while on a leash. Turn their attention to you, and reward them when they keep their attention on you and stay calm around triggers.

2. **Increase Distance:** Gradually move your puppy away from the trigger stimuli until a comfortable threshold is reached. Work on obedience commands and exercises to help your puppy focus from a distance, where he or she can stay calm and listen.

3. **Get help from a professional:** If your puppy's leash reactivity continues or worsens, it's best to talk to a dog trainer or behaviorist with experience working with reactive dogs. They can give you specific strategies and techniques to deal with this behavior problem.

Remember that each puppy is different, and behavior problems may need another solution. During advanced training, if your dog has persistent or severe behavior problems, you should get help from a professional dog trainer specializing in behavior modification. They can give you personalized service and help you deal with specific issues. With patience, consistency, and good ways to solve problems, you can handle behavior problems and move on with your life.

Day 7: Polishing Your Puppy's Skills

Reviewing old lessons:

On the seventh day of caring for your puppy, you should review and reinforce what it has learned over the past week. This step is very important to solidify their training and make sure they remember what they've learned. Review old lessons and build a solid foundation for future training.

Marking the progress of your puppy:

On the last day of caring for your puppy, please take a moment to be proud of how far he or she has come. Recognize how hard you and your furry friend have worked to train each other and how close you have become.

a. **Play and bonding:** Spend time playing with and getting to know your puppy. Use toys, games that get you both involved and gentle touch to get closer to each other. Enjoy their excitement and happiness as they interact with you.

b. **Rewards and Treats:** Reward your puppy for working hard all week with their favorite.

c. **Treats or something special:** Celebrate their achievements and milestones to show how much they have grown and changed.

d. **Reflect and plan:** Take some time to think back on the training process and figure out where your puppy has made the most progress. Think about the next steps in their training and make new goals for them to keep growing. Think about the problems you've solved together and what you've learned from them.

You can end the seven-day puppy care journey on a positive note by reviewing and reinforcing what you've learned so far, easing your puppy into off-leash training, and celebrating his or her progress. Don't forget that training is an ongoing process and that being consistent is important for your puppy's growth. It's important to keep up the training routine and reinforce what they've learned even after the seven days.

As you continue to train your puppy, remember that every day is a chance to grow and improve. Still, they will need consistency, patience, and positive reinforcement to get better at what they do and reinforce good behavior.

Don't forget to change your training methods and plans as your puppy learns more. They may need new challenges and more difficult activities to keep them interested and their minds active. Make the training sessions fit their needs and personalities, and always try to build a relationship based on trust and respect.

Reflecting on Efforts and Your Bond: Consider how hard you and your puppy have worked to train. Think about how much time, patience, and commitment you've both shown this week. Think about the problems you solved together and the big steps you made. Recognize how much your puppy has grown and changed in terms of skills and behavior.

Recognize the trust and connection you and your puppy have built: Appreciate the times when you worked together, talked to each other, and understood each other better. Remember the joy and happiness you all felt when you were training together.

Playing and spending quality time together: To celebrate your puppy's progress, play with it and spend quality time with it. You can play with your puppy using their favorite toys, play games with them, or just give them some physical affection, like a gentle pat on the head or a belly rub. These activities reinforce the good connection between training and having fun, making you and your puppy feel happy and like you're having a good time.

Rewarding and praising: Reward your puppy for working hard and doing well all week long. Use their favorite treats or toys as a way to show them you like them. When your puppy does what you want it to do or does what you tell it to do, praise it enthusiastically by saying things like "Good job!" or "Well done!" Show them how happy you are with how far they've come.

Keeping and Sharing Memories: Use this time to take pictures and write down your memories of training your puppy. Use a camera or your phone to take pictures or videos of them showing off what they have learned. This lets you remember these special times and make lasting memories of your puppy growing up. You can share these memories with your friends and family, or you can keep them to remind yourself of what your puppy has done.

Setting New Goals: Celebrating your puppy's progress is a good time to set new goals for their continued training and development. Think about the ways your puppy can still get better and set realistic goals for the next training phase. Setting new goals helps you keep the momentum going and keeps you challenging your puppy in a good way.

Remember to have fun on the trip with your puppy. Training isn't just about teaching commands and correcting bad behavior. It's also about making a solid bond and giving your furry friend a loving environment. Celebrate their successes and savor the joy of seeing your puppy grow into a well-rounded, happy adult dog.

What should I do if my puppy doesn't respond well to the training methods within the 7-day

Setting Realistic Training Goals:

As you train and help your puppy grow, it's important to set reasonable goals. Setting attainable goals will help you keep track of your puppy's progress, keep yourself motivated, and ensure that the training goes well.

a. **Find Specific Training Areas:** First, determine what specific areas you want to work on with your puppy. It could be basic commands like "sit," "stay," and "come," or it could be specific skills like walking on a leash, going to the bathroom, or getting along with other people. Make a list of the most important skills for your puppy's health and safety.

b. **Separate Goals into Small Steps:** Break each training area into smaller steps that are easier to handle. This method lets you focus on one thing at a time and keeps your puppy from getting too stressed out. For example, if you want to teach your puppy to sit, break it down into steps like luring them with a treat, associating the word "sit" with the action, and slowly taking away the treat.

c. **Set reasonable goals for your puppy's age, breed, and abilities:** Remember that puppies have short attention spans, so training sessions may need to be shorter. Change your goals and expectations accordingly, and only set goals that are manageable and beyond what they can do right now.

d. **Think about the Training Setting:** When setting goals, you should consider the training setting. Start training your puppy in a place that is quiet and free of distractions. Include harder tasks as your puppy gets better. Monitor your puppy's level of activity, focus, and ability to focus. For success, it's important to set goals that are right for the training.

e. **Set a Timeline:** Make a reasonable plan for how long it will take to reach each training goal. Consider your puppy's age, breed, and how quickly it learns. Make sure to set deadlines that are easy to reach because that could lead to frustration or speed up the training process.

f. **Practice often:** Training must be consistent for it to work. Set up a regular schedule for training that lets you practice often. Most of the time, a few short, focused training sessions spread out throughout the day are better than one long session. Regular practice helps your puppy remember what he or she has learned.

g. **Set up ways to measure and keep track of your puppy's progress:** Keep a training journal or use a training app to keep track of your successes, problems, and ways to improve. Keeping track of progress helps you find patterns, change your training methods, and celebrate essential steps.

h. **Adapt and change:** Be flexible and ready to change your goals when needed. Check on your puppy's progress and make any necessary changes to the training plan. If your puppy is having trouble with a specific goal, break it down or ask a professional trainer for help.

i. **Celebrate Successes:** Celebrate your puppy's successes to keep him or her motivated and to reinforce good behavior. Give verbal praise, treats, or time to play as rewards for meeting goals. Celebrating successes makes training more fun for you and your puppy and strengthens your bond.

j. **Seek Professional Help:** If you need help setting realistic goals or encountering problems while training, feel free to ask for help. A certified dog trainer or behaviorist can give you expert advice tailored to your puppy's needs and help you set the right goals.

Problem-Solving and troubleshooting advanced training:

As you advance in your advanced training, you may encounter problems or challenges. Here are some ways to solve problems and fix problems:

1. **Assess the situation:** Find the specific problem or challenge you're having with training your puppy.

2. **Identify the command or behavior:** Find out what command or behavior your puppy is having trouble with. It could be something like "sit," "stay," or "come," or it could be a trick like "rollover" or "shake paw." Knowing the exact area of difficulty will help you train best for that area.

3. **Observe your puppy's response:** During training, pay close attention to how your puppy acts. Do they look confused, bored, or uncertain? Are there signs that they are scared or

worried? Seeing how they move and how they react to commands can give you important clues about what's going on.

4. **Evaluate external factors:** Think about any outside things or distractions that might be getting in the way of your puppy's progress. Are there loud sounds, other pets, or people nearby that might be distracting them? By figuring out these things, you can change the training environment and gradually add distractions as your puppy gets better.

5. **Check the methods and techniques of training:** Think about the ways and methods you've been using to train. Do you use things like treats, praise, and rewards to teach your puppy good things? Are you consistent, and do you give clear, direct instructions? It's important to think about whether your training methods match your puppy's way of learning and personality.

6. **Think about your puppy's age and growth:** Keep in mind that puppies go through different stages of development, and this can change how well they can learn and remember things. Younger puppies may have a shorter ability to pay attention, so they may need to be trained more often. Based on your puppy's age and stage of development, you should change your training goals and methods.

7. **Advice from a qualified dog trainer:** Get help and advice from a qualified dog trainer or behaviorist if you're having trouble with your puppy that doesn't seem to go away or if you don't know how to handle a specific problem. They can give you expert advice, figure out the real problem, and help you find a solution that fits your puppy's needs.

Building a Strong Bond with Your Puppy:

To have a loving and trusting relationship with your puppy, you need to build a strong bond with him or her. Not only operates a strong bond improve your puppy's health as a whole, but it also makes training easier and more fun for both of you. Here is a detailed explanation of how to get close to your puppy:

Spend Quality Time Together: Devote dedicated time each day to spend with your puppy. Perform actions that involve touch, like softly petting, cuddling, and getting a massage. This physical touch makes you feel closer to each other and strengthens your bond.

Create positive associations: Your puppy should think of you as something good. Reward good behavior with treats, praise, and other rewards. Ensure your puppy considers you a happy, safe, and fun place.

Use positive reinforcement: This is a powerful way to build a strong relationship. Give your puppy treats, verbal praise, and affection when it does something good. This reinforces their good behavior and makes your relationship with them stronger.

Communicate Clearly:

1. Talk to your puppy clearly and consistently.

2. Use clear commands and cues that don't change.

3. Pay attention to how your puppy moves and react in the right way.

This makes it easier for people to trust you and helps you communicate better.
Playing together is a great way to get to know each other. Use toys, fetch, tug-of-war, and other games that your puppy likes to play to have fun with it. These activities not only get people moving, but they also bring people together and give them something in common.

1. **Training and Brain Games:** Training sessions don't just teach commands; they also strengthen the bond between you and your pet. Train your puppy using positive reinforcement to teach it new skills and tricks. Use brain games and puzzle toys to challenge their minds, keep them busy, and strengthen your relationship with them.

2. **Be patient and understanding:** It takes time and patience to build a strong relationship. Know that your puppy is still growing and learning. Be patient with your puppy as you train it, and let it make mistakes. Use praise and gentle direction to help them learn and understand what you want from them.

3. **Maintain a Routine:** Make sure your puppy has the same daily routine. Dogs prefer predictable and steady environments. Stick to regular times for eating, playing, exercising, and resting. This routine gives them stability and security, which helps them feel close to each other.

4. **Socialize Your Puppy:** Taking your puppy to different places, meeting new people, and letting it play with other animals helps them feel more confident and builds its bond with you. Set up play dates and trips to the park and supervised meetings with friendly people and well-behaved dogs.

5. **Love and care:** This is the most important thing you can do. Give them a safe, caring place to live and take care of their physical, emotional, and social needs. Your love and care for them over time make a deep and lasting bond.

Remember that building a strong bond with your puppy takes time, patience, and positive reinforcement. By spending quality time with your furry friend, talking to them well, playing with them, training them, and loving them no matter what, you will build a solid and meaningful bond that will last a lifetime.

Building Confidence in Your Puppy:

STAND

Building your puppy's confidence is essential for its health and growth. A puppy that is sure of itself is stronger, more flexible, and better able to handle new situations and challenges.
Here is a full explanation of how to make your puppy more confident:

Make a safe and happy place to work: Give your puppy a place to explore that is both safe and fun. Get rid of any possible dangers or things that might cause fear. Make sure they have a safe, comfortable place to go if they start to feel overwhelmed. Having good things happen around them helps them feel more confident.

Positive Reinforcement Training: To train your puppy, use techniques that give him or her good things. Give them treats, praise, and affection when they do what you want. Focus on building a strong base of basic commands and move on to more difficult tasks slowly. Training sessions keep your puppy's mind active and help to feel better.

Give Your Puppy Chances to Succeed: Set your puppy up for success by giving them challenges that they can handle. Start with easy tasks and make them more complex over time. Honor their achievements and thank them for their hard work. Small wins help puppy feel more confident and like they can do anything.

Exposing Your Puppy to New Places Gradually: Introduce your puppy to new places in a controlled and gradual way. Start with their well-known places, then move on to ones that are slightly different. For example, start in a quiet room in your house and then slowly take them outside, to a park, or to a busy place. Gradually exposing your puppy to new things helps him feel more at ease and builds his confidence.

Encourage Exploration: Let your puppy discover its surroundings at its own pace. Watch them make sure they are safe, but let them explore and touch new things and surfaces on their own. Getting kids to be curious and try new things helps them to gain confidence and a sense of mastery.

Don't be too protective: It's important to ensure the environment is safe but don't be too protective. Letting your puppy face small problems and figure things out independently helps them become strong and confident. To prevent them from getting too busy, keep an eye on how they interact with one another and only intervene when necessary.

Be calm and helpful: Puppies can understand how you feel, so it's important to stay calm and helpful. If you seem sure of yourself and calm, your puppy is more likely to feel safe and sure of itself. Don't react with fear or worry in a new or difficult situation.

Physical and Mental Stimulation: Ensure your puppy gets a lot of physical exercise and mental stimulation. Regular exercise helps your puppy to eliminate pent-up energy and makes their feel better overall. Puzzle toys, interactive games, and training activities that challenge puppy's mind help them to learn how to solve problems and boost confidence.

Patience and consistency: It take time for a puppy to gain confidence, so be patient with his or her progress. Training and socialization efforts need to be consistent for them to work. Use positive reinforcement techniques, keep up with regular routines, and introduce new things slowly. The more consistent you are, your puppy will trust you and feel safe.

You can help your puppy build confidence and resiliency by using these strategies and giving it a supportive and happy environment. Getting them to believe in themselves early on sets them up for growth and success in the future. Remember to be patient, celebrate successes, and give support and guidance.

It's important to remember that every dog is different, and they may learn at different speeds. Some dogs may need more time or special training to deal with certain behaviors or problems. If you talk to a professional dog trainer or take puppy training classes, you can get advice and help tailored to your puppy's specific needs. Remember that the time and money you put into training your puppy will pay off in the form of a well-mannered, happy, and confident puppy that will bring you joy for years to come.

The benefits of puppy training in 7 days and long-term impact on dog's behavior

In just 7 days, training your puppy can have a lot of benefits, and it can also change the way your puppy acts in the long run. Here are some of the best reasons to train your puppy and how it can change the way your puppy acts:

Setting a Strong Foundation: Training your puppy when they are young helps set a strong foundation for how they will act in the future. You can help them do well as they get older by teaching them basic commands and how to act.

Communication and bonding: Training sessions provide you and your puppy a chance to communicate with each other openly. You can build a strong bond based on trust and understanding by praising and rewarding. This bond will keep getting stronger over time, making your relationship with your puppy better as a whole.

Socialization: When you train your puppy, there are often socialization exercises that help your puppy get used to different places, people, and animals. This allows them to feel at ease and act well in different social situations.

Obedience and Control: Teaching your puppy basic commands like "sit," "stay," and "come" helps you get a handle on how they act. It gives them guidance and structure and lets you keep them safe in different situations.

Problem prevention: If unwanted behaviors are addressed and fixed early on through training, they are less likely to become habits. This proactive approach makes it less likely that there will be behavior problems in the future, which will save time and effort in the long run.

Stimulating the mind: Training exercises challenge your puppy's mind and keep it active and stimulated. This can help keep them from getting bored and doing bad things because they have too much energy or nothing to do.

Confidence and Independence: Your puppy's confidence grows as it learns new skills and does well in training exercises. This confidence stays with them even after training is over. It improves their overall attitude and makes it easier for them to get around independently.

Lifelong Learning: The training principles and methods you teach your puppy in the first week will set him up to learn for the rest of his life. Puppies can be trained and given new things to do for the rest of their lives. By starting early, you can raise a puppy open to learning and growing in the future.

Better Adaptation to New Environments: Dogs trained well are better able to adjust to new places and situations. They learn how to stay calm and focused even when they are in a strange place, or there is a lot going on around them. This makes it easier for you to include your puppy in different activities and trips throughout their life.

Better Safety: Training your puppy is a big part of keeping them safe. When they know how to respond to commands like "come" and "stay," they won't run into dangerous situations or do things that could hurt them. This level of control and responsiveness is essential to their safety, especially when they are outside or near busy roads.

Less Anxiety and Stress: Puppies who are trained regularly tend to be less anxious and stressed. Training lets them know what is expected of them, which gives them a sense of safety and predictability. This reduces nervous behaviors like barking too much, chewing things up, or being mean.

Better health and well-being: Training doesn't just change your puppy's behavior; it can also improve his or her physical and mental health. Training sessions give them physical activity, mental stimulation, and a healthy way to eliminate excess energy. This improves their overall fitness and lowers the risk of obesity or other health problems that come with it.

Positive Interactions with Others: A dog trained well is likely to get along better with other dogs, animals, and people. Proper training teaches your puppy good manners and social skills, so he or she can play and interact with other dogs and people without getting angry or acting scared.

Better Control in Emergencies: A well-trained puppy can be essential in an emergency. Whether it's a medical emergency or a potentially dangerous situation, a puppy or dog that reliably follows commands can be easier to handle and keep safe until professional help arrives.

Long-lasting Relationship: Puppy training is good for more than just the puppy's behavior it also helps you and your puppy have a strong, long-lasting relationship. By working hard, you can build a foundation of trust, respect, and mutual understanding that will grow stronger over time.

Quick and Effective Puppy Grooming Training in 7 Days

Most of the time, "grooming" a puppy means cleaning, taking care of, and improving how it appears. Grooming includes performing different things that are good for the puppy's health, hygiene, and well-being.

Day 1: Getting started and brushing up

When you bring your cute new puppy home for the first time, it's important to start grooming them right away and get them used to being touched. This early introduction to grooming sets the stage for a lifetime of good grooming experiences. Here is a more detailed plan for how to go about this:

1. **Make the grooming area calm and inviting:** Find a place in your home that is quiet and comfortable where you can groom your puppy without being interrupted. Make sure the room is warm and well-lit so that it is a nice place for you and your pet.

2. **Gather the tools you'll need:** Before you start, make sure you have the right tools for your puppy's coat type. You can use a slicker brush, a comb, or other tools, depending on the needs of the breed. Talk to an expert or do some research to find out what tools are best for your puppy.

3. **Slowly introduce the grooming tools:** Let your puppy play with the grooming tools and get used to them before you start brushing. Let them smell them, touch them, and even play with them. This step helps them connect the tools to good memories and lessens any fear they might have.

4. **Start with a gentle brushing:** Use slow, gentle strokes with the brush or comb to start grooming. Start with the back or sides, which are less sensitive, and work your way up to the belly and legs, which are more sensitive. Make sure your puppy is comfortable and doesn't pull or tug on their hair. This gentle brushing gets rid of loose hair, stops tangles, and helps keep the coat healthy.

5. **Use positive reinforcement:** During the grooming session, reward your puppy's cooperation and good behavior with words of praise, gentle petting, and small treats. Positive reinforcement helps your puppy learn to like being groomed and makes him or her feel more comfortable during the process.

6. **Take breaks and reassure your dog:** If your puppy shows signs of anxiety or pain, like being restless or making noise, take breaks as needed. Give your puppy a sense of security by talking to it in a calm tone and giving it gentle strokes. As your puppy gets more comfortable and calmer, you can start to groom it for longer periods of time.

7. **Pay attention to body language:** During grooming, it's important to watch your puppy's body language. Look for signs of stress, like flattened ears, a tense body position, or trying to avoid things. If your puppy shows signs of being upset, stop grooming them and let them calm down before going on. If you push them past their comfort zone, they might start to dislike grooming.

8. **Set up a routine for grooming:** When it comes to grooming, it's important to be consistent. Aim for grooming sessions once or twice a week, if possible, to keep your

puppy's coat healthy and looking good. Setting up a routine for grooming your pet not only keeps their coat from getting tangled but also strengthens the bond between you and your furry friend.

Day 2: Bathing

Making sure your new puppy's bathing area is warm and comfortable. Here's a more detailed guide on how to safely and effectively bathe your puppy:

Setting up the bathing area: Make sure the room is warm enough before you bring your puppy in. To make the room feel cozy, you can change the temperature or use a space heater. Put a non-slip mat or towel on the floor of the area where your puppy will bathe to keep it from falling.

Get the things you need for a bath: Prepare all the things you'll need before swimming. This includes a gentle shampoo made just for puppies, a bucket or handheld showerhead for rinsing, soft towels for drying, and any other tools you might need, like a comb or brush.

Choose a puppy-friendly shampoo: Choose a shampoo that is safe for puppies. It is important to use a shampoo that is made for puppies. Normal shampoos made for humans can be too harsh for their sensitive skin and cause irritation. Look for a shampoo made for puppies that are mild and won't cause allergies. If your puppy has a skin problem, talk to your vet about the best shampoo to use.

Gentle introduction to water: Before you bath your puppy, let it explore the bathing area and get used to the sound and feel of water. This makes them feel better and lessens their worry. Use warm water, about the same temperature as their body, to make sure your puppy is comfortable while you bathe it.

Wet and soap up your puppy: Slowly pour the warm water over your puppy's body, starting at its back and working your way down to its tail and legs. Be careful around their face, ears, and eyes because you don't want to get water in those places. Put a small amount of the puppy shampoo on your hands and gently rub it into their fur, making sure to cover all areas. Make sure shampoo doesn't get in their eyes or ears.

Rinse well: After making foam, rinse your puppy well to get rid of all the shampoo. Carefully rinse their fur with the bucket or handheld showerhead, making sure the water flows slowly. Take extra care around their face, and use your hand to keep the water out of their eyes and ears. Remaining shampoo residue can irritate your puppy's skin, so make sure to thoroughly rinse them.

Towel drying and keeping them warm: After you've rinsed your puppy, use a soft towel to dry it off gently. If you don't want to tangle or hurt their coat, pat it instead of rubbing it hard. You can also use a low-heat hair dryer to assist in drying them but be careful not to use too much heat or get too close to their sensitive skin. Wrap your puppy in a towel or use a warm, cozy area to keep them warm during and after drying, especially if it's a cooler place.

Positive reinforcement and praise: While you're bathing your puppy, give it lots of verbal praise, gentle strokes, and treats to reward its good behavior and willingness to help. Positive reinforcement helps your puppy think of bathing well and makes him or her feel more relaxed and at ease during future baths.

Day 3: Nail Trimming

When you trim your puppy's nails, it's important to have the right tools and know what you're doing so that the process is safe and doesn't cause any stress. Here's a more detailed guide on how to trim your puppy's nails:

1. **Choosing the right nail clipper:** It is very important to choose a high-quality nail clipper made just for dogs. There are different kinds, like clippers that look like scissors or clippers that look like guillotine. Do some research or talk to a professional groomer or vet to find out which type will work best for the size and thickness of your puppy's nails.

2. **Learn how the nail clipper works:** Before you try to cut your puppy's nails, take some time to learn how the nail clipper works. Practice holding it and using it to get used to how it works. When it's time to cut your puppy's nails, you'll feel more confident and safer after reading this.

3. **Learn how the nail works:** It's important to know how a dog's nail works so you don't cut into the quick. The quick is a delicate part of the nail that has nerves and blood vessels. If you cut into the vein, your puppy could bleed and feel pain. Take note of how the quick looks on your puppy's nails. The length of the quick can vary depending on the breed and the dog.

4. **Gradual desensitization:** Get your puppy used to having their paws touched before you start cutting their nails. Start by touching their paws gently and massaging them for short periods of time. Over time, you should slowly lengthen these handling sessions. This process helps your puppy get used to having their paws touched and gets them ready to have their nails trimmed.

5. **Position:** Find a place that is comfortable and well-lit to trim your nails. Make sure your dog feels safe and at ease. During the process, you may want to ask a friend or family member to help you hold your puppy gently but securely.

6. **Trim in small steps:** When you trim your nails, you should do it in small steps so you don't cut them quickly. Start by cutting off just a small part of the tip of the nail and work your way back. If your puppy gets stressed or antsy, be careful and take a break. It's better to cut a little at a time than to cut too much at once.

7. **Look at the color and texture of the nails:** When you trim your puppy's nails, pay attention to how they look and feel. Light-colored nails make it easier to see the quick, which looks pink. If your nails are darker, you'll have to look at the texture of the nail to figure out how much to trim. Smooth, see-through nails usually mean you are getting close to the quick, while rough, opaque areas mean you have more space to trim.

Day 4: Clean ears

Cleaning your puppy's ears is an important part of their grooming routine to keep their ears healthy and avoid problems. Here is a detailed guide on how to safely and effectively clean your puppy's ears:

Gather the things you'll need: Before you start, get everything you'll need. This includes cotton balls or pads, a gentle solution made just for dogs to clean their ears, and a towel to catch any drips or extra solution.

Choose an ear-cleaning product made just for dogs: It is important to use a solution made for dogs to clean their ears. Don't use human ear cleaning solutions or products with alcohol on your puppy's ear canal because they can irritate it.

Find a quiet, well-lit place in your home to clean your puppy's ears: This will help you and your puppy feel more at ease by reducing the number of distractions.

Gently look in the ears: Before you start cleaning your puppy's ears, take a moment to look in them. Look for any dirt, discharge, redness, or swelling that you can see. Before you clean your puppy, you should talk to your vet if you notice anything unusual or if your puppy seems to be in pain or discomfort.

Use the ear-cleaning solution: Follow the directions on the bottle and on a cotton ball or pad, put a few drops of the solution. Make sure the cotton ball is damp but not soaked. Too much moisture can make your ears hurt or even give you an ear infection.

Clean the parts of the ear that can be seen: Carefully hold your puppy's ear flap (pinna) and clean the parts of the ear that can be seen, such as the outer ear and the opening to the ear canal. Use the wet cotton ball or pad too gently and carefully wipe away any dirt or dust that you can see. Be careful, and don't put anything deep into your ear canal. Doing so can hurt the sensitive structures inside and cause pain or injury.

Watch how your puppy reacts: Look for any signs of pain or discomfort as you clean it. If your puppy pulls away, whimpers, or shows other signs of pain or distress, stop what you're doing and call your vet right away. If you don't want to hurt your puppy's ears, it's best to ask a professional. After cleaning the ears, use a dry cotton ball or pad to carefully wipe away any extra moisture.

Reward and praise: While you are cleaning your puppy's ears, praise him out loud, pet him gently, and give him treats every so often as a reward for his good behavior. This helps create a good feeling about having your ears cleaned, which makes future sessions easier.

Check for signs of ear problems: You should check your puppy's ears regularly for signs of pain, redness, swelling, a bad smell, too much wax buildup, or changes in behavior. If you notice any of these signs or are worried about your puppy's ears, it is best to take him or her to the vet for a full checkup.

Day 5: Brush the puppy's teeth

The best way to take care of your puppy's teeth is to get them used to having their teeth brushed regularly. Here's a detailed plan for getting your puppy used to toothpaste and brushing its teeth:

1. **Choose a soft-bristled toothbrush and toothpaste made for dogs:** It's very important to use toothpaste made just for dogs since human toothpaste can be dangerous for them. In the same way, choose a toothbrush made for dogs that have soft bristles. Don't use hard-bristled brushes or things that weren't made for dogs because they can hurt their gums and teeth.

2. **Introduce toothpaste slowly:** Start by letting your puppy taste a small amount of toothpaste made for dogs. Put a small amount on the tip of your finger and let them lick it off. This helps them get used to the way the toothpaste tastes and feels. Choose a flavor that your puppy likes so that they can enjoy the experience more.

3. **Introduce the toothbrush slowly:** Once your puppy is used to the toothpaste, you can introduce the toothbrush. Start by letting them sniff and play with the toothbrush while you praise and reward them with treats. This helps them think of the toothbrush as something good.

4. **Start with short brushing sessions:** First, brush your puppy's teeth gently for only 30 seconds to a minute. Focus on the outsides of the teeth and work your way slowly toward the insides. After each brushing session, praise your puppy and give it a treat to keep the sessions positive and rewarding.

5. **Use gentle brushing motions:** When you brush your puppy's teeth, use gentle circular or back-and-forth motions along the gum line. Look at the outside and inside of the teeth as well as the back molars. Be careful and don't brush their teeth too hard, as this can hurt or hurt their gums.

6. **Gradually increase the length of the brushing sessions:** As your puppy gets used to the process, gradually increase the length of the brushing sessions. Aim to brush for a total of two to three minutes, making sure to pay attention to all parts of their teeth and gums.

7. **Be consistent:** Brush your puppy's teeth at least twice or three times a week, not every day, to keep them clean and healthy. Good oral hygiene and avoiding dental problems in the future depend on how often you do things.

When you brush your puppy's teeth, look for signs of dental problems, like bad breath, red or swollen gums, a buildup of tartar, or loose teeth. If you have any worries, you should take your pet to the vet to get a professional dental exam and advice.

Day 6: Clean puppies' eyes and face

As part of their grooming routine, your puppy needs to have clear eyes and a clean face. Start by getting a clean, soft cloth or cotton ball ready. Make sure the cloth or cotton ball is soft and doesn't have any rough or abrasive textures that could hurt your puppy's skin.

1. **Use warm water:** Wet the cloth or cotton ball with warm water before wiping your puppy's eyes and face. Ensure the water is warm enough to feel good but not too hot. Too hot water can hurt or harm your puppy's sensitive skin.

2. **Wipe carefully around the eyes and face:** Hold your puppy's head gently and start by wiping around the eyes. Use a clean damp cloth or cotton ball to carefully wipe away any dirt, dust, or tear stains that may have built up in the area. Be careful not to put too much pressure on the sensitive eye area.

3. **Pay attention to the folds and creases:** Some dog breeds, like Bulldogs and Pugs, have folds and wrinkles on their faces that need extra care when cleaning. Make sure to wipe these areas gently, reaching into the folds to get rid of any dirt or moisture that has gotten stuck there.

4. **Be kind and gentle:** Be kind and gentle with your puppy throughout the whole process. Talk to them in a calm voice and reassure them if they seem worried or uncomfortable. Make cleaning as fun as possible by giving them treats or compliments as a reward for their help.

5. **Water or cleaning solutions should not get in your eyes:** Make sure that neither water nor cleaning products get into your puppy's eyes. When you wipe around the eyes, be careful not to touch the eyes directly. If your puppy is especially sensitive or needs to take care of a specific eye problem, you should talk to your vet for the right advice.

While cleaning your puppy's eyes and face, keep an eye out for any signs of eye irritation, redness, excessive tearing, discharge, or changes in behavior. If you see any of these signs, you should take your pet to the vet for a thorough checkup and the proper treatment.

Day 7: The finishing touches and a review

- On this day, consider your puppies' general grooming needs, like trim puppy's fur.
- Check to see if any more mats, tangles, or places need your attention.
- Give your puppy one last brushing to ensure its fur is clean and knot-free.

Training your puppy to stay calm during visits to the vet or veterinary procedures

Teaching your puppy to stay calm during trips to the vet or veterinary procedures is essential for their overall health and can make these experiences less stressful for you and your furry friend.

1. **Start with pleasant memories:** Make your pet feel good about the veterinary clinic or hospital. Take your puppy to the vet for short trips that don't involve any medical procedures. Let them look around the waiting area and meet the staff. If they behave well, give them treats or praise. This helps your puppy think of the vet clinic as a good place, which will make it less anxious in the future.

2. **Handling and touching your puppy often:** Get your puppy used to being handled and touched gently. Touch their paws, ears, mouth, and other sensitive areas often to make them less scared of veterinary exams. Use treats, praise, and rewards during these handling sessions to help the dog associate being handled with good things.

3. **Gradual exposure to veterinary procedures:** You should slowly and carefully introduce your puppy to common veterinary practices. Start with easy things like touching their paws or ears, and then move on to more complicated things like taking their temperature or looking at their teeth. Always give your puppy a treat for being good and making progress.

4. **Use positive reinforcement techniques:** Positive reinforcement is the best way to help your puppy stay calm at the vet. Use treats, praise, and rewards to encourage puppy to be calm and helpful. For example, give your puppy a treat if it sits still in the waiting room or stays still during a vet exam. This helps them connect being calm with getting a reward, which makes them more likely to remain calm on future visits.

5. **Ask a veterinarian or a veterinary technician for help:** Consult a vet or a veterinary technician if you have a puppy that is especially scared or isn't sure how to handle the situation. They can give you advice tailored to your puppy's needs and help you desensitize and train your puppy to stay calm at the vet.

6. **Keep a calm and reassuring attitude:** Your behavior and attitude affect how your puppy reacts to vet visits in a big way. Stay calm, patient, and reassuring as the process goes on. Dogs can sense how people feel, so staying calm can help them feel better. Talk to them quietly and gently touch them to make them feel better during the visit.

Conclusion

Puppy Training Book For Beginners - Train Your Puppy In just 7 days is a great book that meets the needs of dog owners who want to train their puppies in a short amount of time and need a quick but comprehensive guide. The book is well-written and gives readers a valuable resource that shows off the author's knowledge and deep understanding of how dogs act.

One of the best things about this book is that it condenses the most essential training techniques into a seven-day plan. This time-limited approach gives structure and clarity, making it easy for readers to follow a step-by-step plan that maximizes efficiency and effectiveness. By breaking down the training process into manageable daily tasks, the book gives dog owners the resources they need to take their puppies on a life-changing journey that develops a solid basis of discipline, obedience, and good manners.

Another important thing about the book is that it focuses on positive reinforcement. It shows the importance of rewarding good behavior instead of only punishing bad ones. This approach makes puppies excited to learn and strengthens the bond between owner and pet. By giving clear instructions and helpful tips, the book helps people learn how to talk to their furry friends in a way that helps them get along and trust each other.

Moreover, "Puppy Training Book For Beginners - Train Your Puppy In just 7 days" goes beyond teaching obedience commands. The book is very informative, so people who read it will have everything they need to raise a happy, well-behaved dog. The book encourages dog owners to consider how their puppies feel, which makes training them easier and more compassionate.